A Homework Booklet

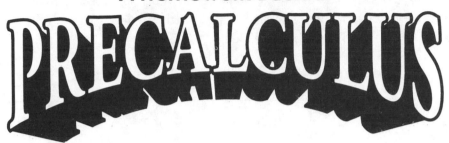

Written by

Wayne J. Bentley

Cover Artist

Vickie Lane

Publishers

Instructional Fair • TS Denison

Grand Rapids, Michigan 49544

Permission to Reproduce

About the Author

Wayne J. Bentley teaches math, government, and debate at a high school for the gifted and talented in Grand Rapids, Michigan. Mr. Bentley earned his bachelor's degree from Grand Valley State University in Allendale, Michigan. Three teams led by Mr. Bentley were state champions in the Academic Quiz Bowl; one team placed in the final four in the 1995 national competition.

Credits

Author: Wayne J. Bentley
Cover Artist: Vickie Lane
Project Director/Editor: Elizabeth Flikkema
Editors: Brad Koch, John Jones
Production: Photo Composition Service

Standard Book Number: 1-56822-418-4
Precalculus
Copyright © 1997 by Instructional Fair • TS Denison
2400 Turner Avenue NW
Grand Rapids, Michigan 49544

Table of Contents

Answer Key in middle of book

Absolute Value Inequalities

> If $a > 0$, then $|x| < a$, only if $-a < x < a$
> $|x| > a$ only if $x < -a$ or $x > a$
> The conclusion of the first formula can also read $-a < x$ and $x < a$.

Example 1: $|3x - 5| < 4$ Solve for x.

Step 1: Immediately write the expression in terms of the definition.
$-4 < 3x - 5$ and $3x - 5 < 4$

Step 2: Solve the algebra. $1 < 3x$ and $3x < 9$
$\frac{1}{3} < x$ and $x < 3$

Step 3: Recombine. $\frac{1}{3} < x < 3$

Step 4: Check. Try a number from the interval in the original equation.
$|3(1) - 5| < 4, |-2| < 4, 2 < 4$. It works.

Example 2: $|3x - 5| \leq 4$ Solve for x.

Follow the steps above except "bring along" the equality sign. Answer: $\frac{1}{3} \leq x \leq 3$

Solve for x. **Note:** Multiplying or dividing by a negative number reverses the inequality.

1. $|5x - 1| < 4$

2. $|2x - 8| > 6$

3. $|5x + 10| < 7 + x$

4. $|4 - 2x| > 6$

The Inverse of a Function

The algorithm for finding the composite inverse of a function is denoted by $f^{-1}(x)$.

Example: Find the composite inverse of the function $f(x) = 3x - 2$ and check the answer.

Step 1: Replace $f(x)$ with y. $y = 3x - 2$

Step 2: Exchange x and y. $x = 3y - 2$

Step 3: Solve the equation in step 2 for y.
$$y = \frac{(x + 2)}{3} = \frac{x}{3} + \frac{2}{3}$$

Step 4: Replace this y with $f^{-1}(x)$. $f^{-1}(x) = \frac{x}{3} + \frac{2}{3}$

Step 5: To check the answer, take the composite of $f(x)$ and $f^{-1}(x)$. The composite of a function and its inverse should equal x.
$$f(f^{-1}(x)) = f\left(\frac{x}{3} + \frac{2}{3}\right) = 3\left(\frac{x}{3} + \frac{2}{3}\right) - 2 = x.$$
Also, $f^{-1}(f(x)) = f^{-1}(3x - 2) = \frac{3x - 2}{3} + \frac{2}{3} = x$

Therefore, the inverse of $f(x) = 3x - 2$ is $f^{-1}(x) = \frac{x}{3} + \frac{2}{3}$.

Find the inverse of the given functions and check the answer.

1. $f(x) = -2x + 7$

2. $f(x) = \frac{1}{x} + 4$

Pascal's Triangle and Expanding a (Binomial)^n

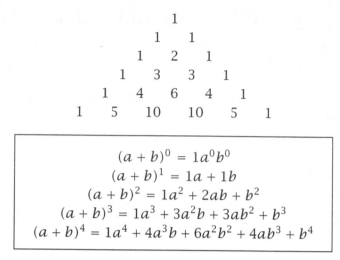

Some general observations on Pascal's Triangle:

1. There is always one more term in the expansion than the power of the binomial.

2. The coefficients of the expansion are derived from Pascal's Triangle.

3. The first term in the binomial occurs in the terms of the expansion in descending powers starting with the power of the binomial and ending with zero.

4. The second term in the binomial occurs in ascending powers in the expansion starting with zero.

Example: Expand the given binomial using Pascal's Triangle.
$$(2x + y)^3$$

Step 1: Find the coefficients of the expansion from Pascal's Triangle.
$+1 + 3 + 3 + 1$

Step 2: Write the first term from the binomial next to the coefficients in descending powers starting with 3. $+1(2x)^3 + 3(2x)^2 + 3(2x) + 1(2x)^0$

Step 3: Write the second term from the binomial behind each term in ascending powers starting with zero. $+1(2x)^3 y^0 + 3(2x)^2 y + 3(2x)y^2 + 1(2x)^0 y^3$

Step 4: In this problem, raise the 2 in each term to its respective power and multiply it by its coefficient. $(2x + y)^3 = 8x^3 + 12x^2 y + 6xy^2 + y^3$

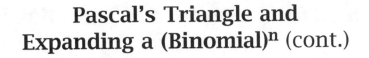

Pascal's Triangle and Expanding a (Binomial)^n (cont.)

1. Add the next row in the coefficient pyramid for Pascal's Triangle. What is the power of the binomial it represents?

2. Expand the binomial $(x - y)^4$. **Hint:** $(x - y)^4 = (x + (-y))^4$

3. Expand the binomial $(a - 3b)^3$.

4. a. What is the leading coefficient in the expansion of the given binomial?
   $(5x + 2y)^6$

   b. What is the last coefficient?

   c. What is the middle coefficient?

5. Another application of Pascal's Triangle is factoring. Factor the following polynomial completely. **Hint:** factor out a constant term and look for the coefficient pattern.

$$P(x) = 7x^5 + 35x^4y + 70x^3y^2 + 70x^2y^3 + 35xy^4 + 7y^5$$

Standard Forms for Conic Sections

General Equation
$Ax^2 + Bxy + Cy^2 + Dx + Ey + F = 0 \qquad B = 0$
Standard Forms
Parabola: $(y - k) = a(x - h)^2$ or $(x - h) = a(y - k)^2$
Ellipse: $\dfrac{(x - h)^2}{a^2} + \dfrac{(y - k)^2}{b^2} = 1$ or $\dfrac{(x - h)^2}{b^2} + \dfrac{(y - k)^2}{a^2} = 1$
Circle: $(x - h)^2 + (y - k)^2 = r^2$
Hyperbola: $\dfrac{(x - h)^2}{a^2} - \dfrac{(y - k)^2}{b^2} = 1$ or $\dfrac{(y - k)^2}{a^2} - \dfrac{(x - h)^2}{b^2} = 1$

Example: Given the general equation for conic sections, convert it to standard form.
$$25x^2 + 9y^2 + 100x - 54y - 44 = 0$$

Step 1: Group x and y terms. $\quad (25x^2 + 100x) + (9y^2 - 54y) - 44 = 0$

Step 2: Complete the square on both x and y.
$$25(x^2 + 4x) + 9(y^2 - 6y) = 44$$
$$25(x^2 + 4x + 4) + 9(y^2 - 6y + 9) = 44 + 100 + 81$$
$$25(x + 2)^2 + 9(y - 3)^2 = 225$$

Step 3: In order to get 1 on the right side of the equation, divide the equation by 225.
$$\frac{25(x + 2)^2}{225} + \frac{9(y - 3)^2}{225} = \frac{225}{225}$$

Step 4: Reduce the fractions. $\dfrac{(x + 2)^2}{9} + \dfrac{(y - 3)^2}{25} = 1$

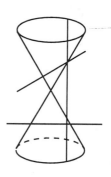

Given the general equation for conic sections, find the standard equation.

1. $x^2 - 4y^2 - 6x - 8y + 9 = 0$ 　　　　　2. $x^2 + y^2 + 8x - 6y + 21 = 0$

Intersecting Graphs

In calculus, students have to find the area between curves and revolve the area around an axis. In order to accomplish this task, they have to be able to find the point(s) of intersection (a precalculus activity).

Substitution Method

Example: Find the points of intersection between the line $y = 2x + 3$ and the parabola $y = x^2$.

Step 1: Substitute the y-value of the line into the y-value of the parabola to obtain an equation in x.
$2x + 3 = x^2$

Step 2: Find the roots of the equation. Use factoring.

$$x^2 - 2x - 3 = 0$$
$$(x + 1)(x - 3) = 0$$
$$x = -1, x = 3$$

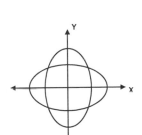

When $x = -1$, $f(x) = 1$, and when $x = 3$, $f(x) = 9$. Therefore, $(-1, 1)$ and $(3, 9)$ are the points of intersection between the line and the parabola.

Find the point(s) of intersection of the two curves.

1. $y = 3x, \quad y = -2x + 5$

2. $y = -7x, \quad y = 2x^2 + 3$

Intersecting Graphs (Conic Sections)

Substitution Method

Example: Find the points of intersection of the two conic section graphs.

$$x^2 + y^2 = 4 \text{ and } x^2 + 3y^2 - 6y = 12$$

Step 1: Solve for x^2 in terms of y^2.

$$x^2 = -y^2 + 4$$

Step 2: Substitute for x^2 into the remaining equation.

$$-y^2 + 4 + 3y^2 - 6y = 12$$

Step 3: Combine and solve for y.

$$2y^2 - 6y - 8 = 0$$
$$y^2 - 3y - 4 = 0$$
$$(y + 1)(y - 4) = 0$$

$y = -1$ or $y = 4$. However, $y \neq 4$ because it is not part of the range of the first equation.

Step 4: Substitute -1 in for y to find the x coordinate.

$$x^2 + (-1)^2 = 4$$
$$x^2 = 3 \text{ and } x = +\sqrt{3} \text{ and } -\sqrt{3}$$

Therefore the points of intersection of the two curves are $(-\sqrt{3}, -1)$ and $(\sqrt{3}, -1)$.

Find the point(s) of intersection of the two conic sections.

1. $x^2 + y^2 = 9$ and $\dfrac{x^2}{16} + \dfrac{y^2}{9} = 1$

2. $x^2 + y^2 = 16$ and $x^2 + y^2 + 4x = 11$

Complete the Square

General Form	Standard Form
$y = ax^2 + bx + c$	$(y - k) = a(x - h)^2$

Write the general equation for a parabola in standard form by completing the square.

Example: $y = 4x^2 + 8x - 6$

Step 1: Bring the constant term over to the left side of the equation.
$y + 6 = 4x^2 + 8x$

Step 2: Factor out the leading coefficient. $y + 6 = 4(x^2 + 2x)$

Step 3: a. Take half the x coefficient. $\dfrac{2}{2} = 1$

b. Square it. $(1)^2 = 1$

c. Put it inside the parentheses. $(x^2 + 2x + 1)$

d. Multiply it by the leading coefficient. $(4 \cdot 1 = 4)$

e. Add the result to the left side of the equation. $y + 6 + 4 = 4(x^2 + 2x + 1)$

Step 4: What is inside the parentheses should be a perfect square.
$(x^2 + 2x + 1) = (x + 1)^2$

Step 5: Simplify, and the equation is in standard form. $y + 10 = 4(x + 1)^2$

Write the general equation for a parabola in standard form by completing the square.

1. $y = 3x^2 - 9x + 5$ 2. $y = -5x^2 + 20x - 3$

3. $y = x^2 + 2 + 3$

Solving Simple Logarithmic Equations
(Change-of-Base Rule)

---

1. $\log_b x = y$ is equivalent to $b^y = x$ where $x, b > 0$ and $b \neq 1$.
2. Change-of-base rule
   $$\log_b x = \frac{\log_a x}{\log_a b} \text{ where } a, b, \text{ and } x > 0 \text{ and } a, b \neq 1.$$
   Note: $\log x = \log_{10} x$ and $\ln x = \log_e x$

In order to solve a simple logarithmic equation, write it as an exponential equation.

Example 1: Solve for x.

$$\log_3 x = 2$$
$$3^2 = x \qquad \text{Formula 1}$$
$$x = 9$$

Example 2: Solve for x and round to the nearest hundredth.

$$x = \log_2 5$$
$$= \frac{\log_{10} 5}{\log_{10} 2} = \frac{\log 5}{\log 2} = 2.32 \qquad \text{Formula 2}$$
$$\text{(calculator)}$$

Solve for x.

1. $\log_5 x = 2$ 　　　　　2. $\log_3 2x = 7$ 　　　　　3. $\log_4(x - 1) = 2$

Solve for x using the natural log(ln) and round to the nearest hundredth.

4. $x = \log_2 12$ 　　　　　5. $3x = \log_2 32$ 　　　　　6. $\log_x 16 = 4$

Solving Exponential Equations
(Without Logarithms)

Method 1: Solving exponential equations is simplest when the bases are equal. If the bases are equal, then the exponents are equal.

Example:

$(4)^{x+2} = (4)^{2x-2}$

The bases are both 4, therefore

$x + 2 = 2x - 2$ and $x = 4$.

Check the answer. $(4)^{4+2} = (4)^{2(4)-2}$

$$(4)^6 = (4)^6$$

Method 2: If the bases are not equal, but they can be made equal because they are integer powers of the same base, then make the bases equal and the exponents will be equal.

Example:

$(3)^x = (27)^{5-2x}$

$27 = (3)^3$

Substitute in the new base.

$(3)^x = ((3)^3)^{5-2x}$

$(3)^x = (3)^{15-6x}$

The bases are both 3, therefore

$x = 15 - 6x$

$7x = 15$

$x = \dfrac{15}{7}$

Solve the following exponential equations.

1. $2^{2x+1} = 2^{x-3}$

2. $5^{2-2x} = 5^{x+2}$

3. $10^2 = 10^{x-4}$

4. $9^{x+7} = 3^{3-x}$

Solving Exponential Equations with Logs

$$\text{If } x = y, \text{ then}$$
$$\log_b x = \log_b y$$
$$\text{where}$$
$$b, x, \text{ and } y > 0 \text{ and}$$
$$b \neq 1$$

Solve for x and round to the nearest thousandth.

Example 1: $\left(10^{2x}\right)^2 = 3^{x+1}$

Step 1: Simplify the exponent.
$$10^{4x} = 3^{x+1}$$

Step 2: Take the log of both sides.
$$\log 10^{4x} = \log 3^{x+1}$$

Step 3: Take the exponent out front of the log.
$$4x \log 10 = (x + 1) \log 3$$

Step 4: Use the $\log 10 = 1$ and distribute.
$$4x = x \log 3 + 1 \log 3$$

Step 5: Group the terms with an x on one side and factor out the x.
$$x(4 - \log 3) = \log 3$$

Step 6: Solve.
$$x = \frac{\log 3}{(4 - \log 3)} = .135$$

Example 2:

$$12^x = 3^{2x-1}$$
$$\log 12^x = \log 3^{2x-1}$$
$$x \log 12 = (2x - 1) \log 3$$
$$x \log 12 = 2x \log 3 - \log 3$$
$$\log 3 = 2x \log 3 - x \log 12$$
$$\log 3 = x(2 \log 3 - \log 12)$$
$$x = \frac{\log 3}{(2 \log 3 - \log 12)} = -3.189$$

Solving Exponential Equations with Logs (cont.)

Solve for x to nearest thousandth.

1. $5^{x+2} = 10^{2x}$

2. $7^x = 10^{2x-5}$

3. $10^x = 6^{x-2}$

4. $7^{2x} = 3^{5x-2}$

5. $5^{3x} = 8^{2x-3}$

6. $9^{2x} = 3^{-x+7}$

7. $12^x = 6^{2x-2}$

8. $11^{2x} = 102^{x-1}$

Compound Interest

$$A = P\left(1 + \frac{r}{n}\right)^{nt}$$

A is the amount in the account after the interest is added.
P is the original principal amount in the account.
r is the yearly interest rate, expressed as a decimal, at which the account increases.
t is the length of time in years that the interest accumulates.
n is the number of times compounded annually.

Example 1: Find the amount in the account when the principal is $1000, compounded quarterly, for three years, at 6% interest.

Step 1: Identify each variable using a ? for the variable you have to find.

$\quad A = ? \quad P = \$1000 \quad r = .06 \quad t = 3 \quad n = 4$

Step 2: Substitute into the equation.

$$A = 1000\left(1 + \frac{.06}{4}\right)^{(.06)3}$$

Step 3: Solve the equation for the variable you need to find.
(When money, round to the nearest cent.) $A = \$1,002.68$

Example 2: How long would it take for $2,000 to double if 5% interest is compounded semi-annually?

Step 1: $A = \$4,000 \quad P = \$2000 \quad r = .05 \quad t = ? \quad n = 2$

Step 2: $4,000 = 2,000\left(1 + \dfrac{.05}{2}\right)^{(2)t}$

Step 3: a. Divide both sides by 2,000 and combine the numbers inside the parentheses. $2 = (1.025)^{(2)t}$

b. Take the common log of both sides. $\log 2 = \log(1.025)^{(2)t}$

c. Using the log combination rule for exponents, bring the exponent out front. $\log 2 = 2t\log(1.025)$

d. Therefore: $t = \log 2/(2\log(1.025))t \approx 14$ years
Note: If $\dfrac{r}{n}$ doesn't come out exact when dividing, then combine $1 + \dfrac{r}{n}$ as $\dfrac{n+r}{n}$ for best accuracy.

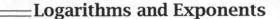

Compound Interest (cont.)

$$A = P\left(1 + \frac{r}{n}\right)^{nt}$$

Solve for the missing variable in each problem.

1. $A = \$2000$ $P = \$800$ $r = 3.5\%$ $n = 12$ and $t =$? years (nearest tenth)

2. $A = \$2500$ $P = $? $r = 4.5\%$ $n = 6$ $t = 4$ years

3. $A = $? $P = \$2000$ $r = 4\%$ $n = 6$ $t = 10$ years

4. $A = \$4000$ $P = \$2000$ $r = 4\%$ $n = 6$ $t = $? years (nearest tenth)

Continuous Compound Interest

$$A = Pe^{rt}$$

A is the amount in the account after the continuous interest is added.
P is the original principal amount in the account.
r is the yearly interest rate, expressed as a decimal, at which the account increases.
t is the length of time in years that the interest is accumulated.
$e \approx 2.7182818284530$ (the shift ln key on a scientific calculator)

Example 1: Find the amount in the account when the principal is $1000, compounded continuously, for three years, at 6% interest.

Step 1: Identify each variable using a *?* for the variable you have to find.
   $A = ?$ $P = \$1000$ $r = .06$ $t = 3$

Step 2: Substitute into the equation.
   $A = 1000e^{(.06)3}$

Step 3: Solve the equation for the variable you have to find.
   $A = \$1197.22$

Example 2: How long would it take for $2,000 to double if 5% interest is compounded continuously? Round to the nearest tenth.

Step 1: $A = \$4,000$ $P = \$2000$ $r = .05$ $t = ?$

Step 2: $4,000 = 2,000e^{(.05)t}$

Step 3: a. Divide both sides by 2,000. $2 = e^{(.05)t}$

   b. Take the natural log of both sides. $\ln 2 = \ln e^{(.05)t}$

   c. Using the log combination rule for exponents, bring the exponent out front. $\ln 2 = .05t \ln e$

   d. Remember the $\ln e = 1$, therefore: $t = \dfrac{\ln 2}{.05}$
   $$t = 13.9 \text{ years}$$

Continuous Compound Interest (cont.)

Solve for the missing variable in each problem.

1. A = \$1200 P = \$300 r = ? t = 4 years (nearest hundredth percent)

2. A = ? P = \$2700 r = 3.5% t = 3 years

3. A = \$800 P = ? r = 6% t = 10 years

4. A = \$1800 P = \$1350 r = 6.2% t = ? years (nearest hundredth)

Continuous Growth and Radioactive Decay

$$A = A_o e^{kt}$$

Continuous growth and radioactive decay work just like continuous compound interest. The amount of material at the end of a time frame is dependent on the amount with which you start and the rate at which it changes. The rate at which the material grows or decays, k, will be a negative value if the amount is decreasing, and positive if the amount is increasing. A_o is the original amount of material.

Example 1: At what rate is the material decaying if, after 150 years, 100 grams remain of a 200-gram sample? **Note:** 150 years is the half-life of the material. (Round to 4 decimal places.)

Step 1: Identify each variable. $A_o = 200$ grams $A = 100$ grams $t = 150$ years $k = ?$

Step 2: Substitute into the formula. $100 = 200e^{k150}$

Step 3: Solve for the unknown variable.

a. Divide both sides by 200. $.5 = e^{k150}$

b. Take the natural log of both sides. $\ln.5 = \ln e^{k150}$

c. Using the log combination rules for exponents, bring the exponent out front. $= \ln.5 = (150k)\ln e$

d. $\ln e = 1$, therefore, $k = \dfrac{(\ln.5)}{150} \approx -.0046$

Example 2: At what rate is the amount of material growing if 100 bacteria in a sample increase to 250 bacteria when the sample is observed for 150 minutes? (Round to 4 decimal places.)

Step 1: Identify each variable. $A_o = 100$ bacteria $A = 250$ bacteria $t = 150$ minutes $k = ?$

Step 2: Substitute into the formula. $250 = 100e^{k150}$

Step 3: Solve for the unknown variable.

a. Divide both sides by 100. $2.5 = e^{k150}$

b. Take the natural log of both sides. $\ln 2.5 = \ln e^{k150}$

c. Using the log combination rules for exponents, bring the exponent out front. $\ln 2.5 = (150k)\ln e$

d. $\ln e = 1$, therefore, $k = \dfrac{(\ln 2.5)}{150} \approx .0061$

Continuous Growth and Radioactive Decay (cont.)

Solve for the rate of growth or decay. Round to the number of places indicated.

1. What is the rate of decay if 250 grams remain of a material that has a half-life of 2700 years? (4 decimal places)

2. What is the rate of growth of 10 bacteria if after 3.5 hours there are 2,000 bacteria? (4 decimal places)

3. How long will it take 50 grams of a substance to decay to 10 grams if the rate of decay is $k = -.345$? (nearest tenth)

4. How many bacteria will there be if 100 bacteria increase at a rate of $k = 2.5$/minute for 10 hours? (two significant figures)

Summation Notation and Arithmetic Series

An example of summation notation is $\sum_{k=2}^{5} (3k - 1)$.

This is read, "The sum of $3k - 1$ as k goes from 2 to 5." $\sum$ is a Greek symbol called sigma which means *sum*. k is called the index of summation. Any other letter could be used.

Example: In order to write $\sum_{k=2}^{5} (3k - 1)$ in expanded form, substitute whole numbers from 2 to 5 in for k and add as shown.

$$\sum_{k=2}^{5} (3k - 1) = (3k - 1) = 3(2) - 1 + 3(3) - 1 + 3(4) - 1 + 3(5) - 1 = 38$$

Expand and find the sum.

1. $\sum_{i=4}^{8} (i)^2 =$

2. $\sum_{p=2}^{6} 3^p =$

3. $\sum_{t=2}^{7} 10 - t =$

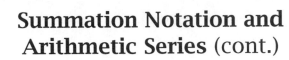

Summation Notation and
Arithmetic Series (cont.)

An *arithmetic series* is an expanded sum of numbers such that each pair of consecutive numbers has a common difference. An example of this is $2 + 6 + 10 + 14 + 18 + \ldots$, where the common difference is $4 (6 - 2 = 4, 10 - 6 = 4,$ etc.). To find the nth term of an arithmetic series, use the following formula: $t_n = a_1 + (n - 1)d$; where $t_n = n^{th}$ term, $a =$ first term, $n =$ number of terms, and $d =$ common difference. To find the 6th term of the above series, do this:

$$t_n = a_1 + (n - 1)d$$
$$t_6 = 2 + (6 - 1)4 = 22$$

Now, place 22 in the series and see if it fits.

Write the following arithmetic series in summation notation ($\sum$).

Example: $-2 + 1 + 4 + 7 + 10 + 13$

Step 1: Write $\sum\limits_{n=1}^{6}$ (There are 6 terms in this series; number them from 1-6.)

Step 2: $l_n = a_1 + (n - 1)d$
$$t_n = -2 + (n - 1)3$$
$$= 3n - 5$$

So write $\sum\limits_{n=1}^{6} (3n - 5)$

Step 3: Check (optional).

$$\sum_{n=1}^{6} (3n - 5) = 3(1) - 5 + 3(2) - 5 + 3(3) - 5 + 3(4) - 5 + 3(5) - 5 + 3(6) - 5$$
$$= -2 + 1 + 4 + 7 + 10 + 13$$

1. $3 + 8 + 13 + 18$

2. $(-5) + (-1) + 3 + 7$

Geometric Series

A *geometric series* is an expanded sum of numbers (terms) such that there is a common ratio between any two consecutive terms. An example is $1 + 3 + 9 + 27 + \ldots$ where the common ratio is 3 ($\frac{3}{1} = 3$, $\frac{9}{3} = 3$, etc.). Use the formula $s_n = \dfrac{a_1(1 - r^n)}{1 - r}$ to find the sum of the first n terms where $s_n =$ the sum of the first n terms, $a_1 =$ first term, $r =$ the common ratio, and $n =$ the number of terms.

Find the sum of the finite geometric series.

Example: $1 + \dfrac{1}{2} + \dfrac{1}{4} + \ldots + \dfrac{1}{256}$

Step 1: Notice the common ratio is $\dfrac{1}{2}$. Fill in the missing terms and see that

there are 9 terms. $1 + \dfrac{1}{2} + \dfrac{1}{4} + \dfrac{1}{8} + \dfrac{1}{16} + \dfrac{1}{32} + \dfrac{1}{64} + \dfrac{1}{128} + \dfrac{1}{256}$

Step 2: Use $s_n = \dfrac{a_1(1 - r^n)}{1 - r}$ $\qquad s_9 = \dfrac{1(1 - \frac{1}{2}^9)}{1 - \frac{1}{2}} = \dfrac{511}{256}$

1. $3 + 6 + 12 + 24 + 48 + \ldots + 384$

2. $\dfrac{1}{2} + \dfrac{1}{4} + \dfrac{1}{8} + \dfrac{1}{16} + \dfrac{1}{32}$

Geometric Series (cont.)

An infinite geometric series has a sum if and only if $|r| < 1$. Then the following formula may be used to find the sum:

$$s = \frac{a_1}{1-r}$$

Find the sums of the infinite geometric series.

Examples:

a. $2 + 3 + \dfrac{9}{2} + \dfrac{81}{4} + \ldots$

$|r| = \dfrac{3}{2}$ and $\dfrac{3}{2} > 1$ so you can't find the sum.

b. $2 + 1 + \dfrac{1}{2} + \dfrac{1}{4} + \ldots$

$|r| = \dfrac{1}{2}$ so you may use the formula.

$$s = \frac{a_1}{1-r} = \frac{2}{1 - \frac{1}{2}} = 4$$

1. $1 + 2 + 4 + 8 + \ldots$

2. $\dfrac{3}{2} + \dfrac{1}{2} + \dfrac{1}{6} + \dfrac{1}{18} + \ldots$

3. $-2 + \dfrac{2}{3} - \dfrac{2}{9} + \dfrac{2}{27} + \ldots$

Mathematical Induction Recursive Definition

Given the recursive definition for the sequence, follow the steps to prove by mathematical induction that the explicit expression for the nth term generates the terms of the sequence.

The sequence: $1, 4, 13, 40, \ldots, \dfrac{3n-1}{2} \ldots$ The recursive definition: $a_1 = 1$

$$a_{n+1} = 3a_n + 1$$

Step 1: From the statement of the problem, write an equation for the nth term defined by the explicit expression.

Let $S = \left\{ n \in N : a_n = \dfrac{3^n - 1}{2} \right\}$

Step 2: Show $1 \in S$. Then take the first term of the sequence or series and set it equal to the explicit expression with 1 substituted in for n. If it turns out to be equal, then write "true $1 \in S$."

$$1 = \frac{3^1 - 1}{2} = 1 \quad \text{true } 1 \in S$$

Step 3: Assume $x \in S$. Then go back to the equation in Step 1 and substitute x for n.

$$a_x = \frac{3x - 1}{2}$$

Step 4: Prove $x + 1 \in S$. Now go back again to the equation in Step 1 and substitute $(x + 1)$ for n.

$$a_{x+1} = \frac{3^{x+1} - 1}{2}$$

Step 5: Proof. Now go to the recursive definition and write the equation for the $x + 1$ term. Substitute into this equation the x term from the assumption; work with it algebraically until it matches the "prove" statement.

$$a_{x+1} = 3a_x + 1 \qquad \left(\text{From Step 3: } a_x = \frac{3^x - 1}{2}\right)$$

$$= 3\left(\frac{3^x - 1}{2}\right) + 1$$

$$= \frac{3 \cdot 3^x - 3}{2} + \frac{2}{2} = \frac{3^1 \cdot 3^x - 3 + 2}{2} = \frac{3^{x+1} - 1}{2}$$

Step 6: $\therefore S = N$. Therefore, the explicit expression for the nth term generates the same sequence or series as the recursive definition.

Mathematical Induction
Recursive Definition (cont.)

Complete the following exercises.

1. Prove by mathematical induction that $a_n = 2^n + 1$ is the explicit expression for the nth term for the sequence defined recursively by the following:

$a_1 = 3$

$a_{n+1} = 2a_n - 1$

2. Prove by mathematical induction that $a_n = 5(\frac{1}{2})^{n-1}$ is the explicit expression for the nth term for the sequence defined recursively by the following:

$a_1 = 5$

$a_{n+1} = \dfrac{a_n}{2}$

3. Prove by mathematical induction that $a_n = \dfrac{n}{2} - \dfrac{9}{2}$ is the explicit expression for the nth term for the sequence defined recursively by the following:

$a_1 = -4$

$a_{n+1} = a_n + \dfrac{1}{2}$

Mathematical Induction Summation Notation

Example: Prove that $\sum_{i=1}^{n} 2i = n(n + 1)$ is true for all natural numbers.

Step 1: Write out the series equation that corresponds to the $\sum$.

$$\text{Let } S = \left\{ n \in N : \sum_{i=1}^{n} 2i = 2 + 4 + 6 + \ldots + 2n = n(n + 1) \right\}$$

Step 2: Show $1 \in S$. See if the first term is equal to the general formula for the sum of the first n terms of the series, substituting 1 for n. If they are equal, then write "true $1 \in S$."

$1(1 + 1) = 2 \quad$ true $1 \in S$

Step 3: Assume $x \in S$. Copy the series equation replacing n with x.

$2 + 4 + 6 + \ldots + 2x = x(x + 1)$

Step 4: Prove $x + 1 \in S$. Copy the series equation replacing n with $x + 1$. (It is important to include the last 2 terms of the series.)

$2 + 4 + 6 + \ldots + 2x + 2(x + 1) = (x + 1)[(x + 1) + 1]$

Step 5: Proof. This is the pivotal point of mathematical induction; you must find the difference between the left-hand side of the *assumption* equation and the left-hand side of the *prove* statement, then add this difference to both sides of the assumption. The difference is $2(x + 1)$.

$$2 + 4 + 6 + \ldots + 2x + 2(x + 1) = x(x + 1) + 2(x + 1)$$
$$\text{(Add } 2(x + 1) \text{ to both sides of the equation in Step 3.)}$$
$$= x^2 + x + 2x + 1$$
$$= x^2 + 3x + 1$$
$$= (x + 1)(x + 2)$$
$$\text{(Rewrite } x + 2 \text{ as } (x + 1) + 1)$$
$$= (x + 1)[(x + 1) + 1]$$

Step 6: Now work with the right-hand side of this equation and the right-hand side of the prove statement algebraically until they are identical.

Step 7: Therefore $S = N$.

Mathematical Induction
Summation Notation (cont.)

Prove each statement by mathematical induction.

1. $\displaystyle\sum_{i=1}^{n} \frac{1}{i(i+1)} = \frac{n}{(n+1)}$

2. $\displaystyle\sum_{i=1}^{n} i^3 = \frac{[n(n+1)]^2}{4}$

Limits at Infinity

$$F(x) = \frac{Q(x)}{P(x)}$$

$F(x)$ is the quotient between two polynomials.

Rules for taking the limit of $f(x)$ as x approaches infinity:

1. If the degree of the numerator is equal to the degree of the denominator, then the limit is the ratio of the leading coefficients.

2. If the degree of the numerator is less than the degree of the denominator, then the limit is zero.

3. If the degree of the numerator is greater than the degree of the denominator, then the limit as a number does not exist.

Examples:

1. $\lim\limits_{x \to \infty} \dfrac{3x^3 - 2x^2 + 2x - 4}{7x^3 - 3x^2 + 2x - 27} = \dfrac{3}{7}$

   The degree of the numerator (3) = degree of the denominator (3) (rule 1).

2. $\lim\limits_{x \to \infty} \dfrac{4x^2 - 2x + 27}{2x^3 - 5x + 1} = 0$

   The degree of the numerator (2) < degree of the denominator (3) (rule 2).

3. $\lim\limits_{x \to \infty} \dfrac{9x^2 + 7}{x - 2}$ as a number does not exist

   The degree of the numerator (2) > degree of the denominator (1) (rule 3).

Limits at Infinity (cont.)

Find the limit.

1. $\lim\limits_{x\to\infty} \dfrac{3x^2 - 7x + 2}{x^2 - 81} =$

2. $\lim\limits_{x\to\infty} \dfrac{-7x^3 - 2x + 27}{4x^4 - x^2 + 5} =$

3. $\lim\limits_{x\to\infty} \dfrac{-12x^2 - 24}{3x^2 - 6} =$

4. $\lim\limits_{x\to\infty} \dfrac{7x^5 - 3x^3 + 2x}{x^3 - 5x^2 + 3x} =$

5. $\lim\limits_{x\to\infty} \dfrac{x^2 - 5x^3 - 7x^4}{7x^4 - 14x^2 + 1} =$

6. $\lim\limits_{x\to\infty} \dfrac{27 - 3x^2}{3x^2 + 27} =$

AUTOMATIC GRAPHER: Use your automatic grapher to graph one from each rule to observe the behavior as the function goes to infinity.

Points of Discontinuity

$$F(x) = \frac{Q(x)}{P(x)}$$

$F(x)$ is the quotient between two polynomials.

Following is the algorithm for finding points of discontinuity.

Step 1: Given the quotient of two polynomials, take the denominator and set it equal to zero.

Step 2: Factor the equation, if possible.

Step 3: Find the value(s) of x that make the denominator zero.

Step 4: Substitute these values for x into the numerator; if any x makes the numerator zero, then that value may be the x-coordinate of a point of discontinuity.

Step 5: To find the y-coordinate of the point of discontinuity, factor the numerator, cancel like factors with the denominator, then substitute again the x value(s) from step 4. If this substitution makes the denominator 0, then there is no point of discontinuity.

Example:

$$F(x) = \frac{x^2 - 4}{x^2 - 2x - 8}$$

1. $x^2 - 2x - 8 = 0$

2. $(x + 2)(x - 4) = 0$

3. $x = -2, 4$

4. $(-2)^2 - 4 = 0, (4)^2 - 4 \neq 0$
   Therefore -2 is the x-coordinate of a point of discontinuity.

5. $\dfrac{(x - 2)(x + 2)}{(x + 2)(x - 4)} = \dfrac{(x - 2)}{(x - 4)}$,

   Substituting -2 , $\dfrac{(-2 - 2)}{(-2 - 4)} = \dfrac{-4}{-6} = \dfrac{2}{3}$

   Therefore, the point of discontinuity is $(-2, \dfrac{2}{3})$.

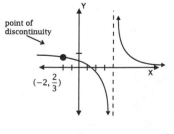

point of discontinuity

$(-2, \frac{2}{3})$

Points of Discontinuity (cont.)

Find the point(s) of discontinuity.

1. $F(x) = \dfrac{x^2 - 1}{x^2 - 5x - 6}$

2. $F(x) = \dfrac{x^2 - 4x - 5}{x^2 - 25}$

3. $F(x) = \dfrac{x^2 + 10x + 9}{x^2 - 81}$

AUTOMATIC GRAPHER: Graph one problem, then use the trace function (or another method) to check your answer.

Horizontal and Oblique Asymptotes

$$F(x) = \frac{Q(x)}{P(x)}$$

$F(x)$ is the quotient between two polynomials.

When the limit of $f(x)$ as $|x|$ approaches infinity is zero or a constant, then $y = $ the limit is a horizontal asymptote. $|x| \to \infty$ means $x \to \pm$

Example 1: $f(x) = \dfrac{x}{(x-1)}$

$\displaystyle\lim_{|x|\to\infty} \frac{x}{(x-1)} = 1$ Therefore, $y = 1$ is an horizontal asymptote.

Example 2: $\displaystyle\lim_{|x|\to\infty} \frac{x}{(x^2-1)} = 0 \therefore y = 0$ is a horizontal asymptote.

When the limit of $f(x)$ as x approaches infinity does not exist as a number and the degree of the numerator is one more than the degree of the denominator, then the quotient of two polynomials will have an oblique asymptote.

Example 3: Find the horizontal or oblique asymptote. $f(x) = \dfrac{(x^2-2)}{(x-1)}$

Step 1: Take the limit as x approaches infinity.

$\displaystyle\lim_{|x|\to\infty} \frac{(x^2-2)}{(x-1)} = \pm\infty$ (does not exist) See case 3 of "Limit of a Function

at a Point" (p. 42).

Step 2: The degree of the numerator is one more than that of the denominator. Therefore, $f(x)$ has an oblique asymptote. To find the oblique asymptote, perform long division on the polynomials.

$$
\begin{array}{r}
x + 1\frac{-1}{(x-1)} \\
\hline
x - 1 \overline{\smash{\big)}\, x^2 + 0x - 2} \\
\underline{x^2 - x} \\
0 + x - 2 \\
\underline{+ x - 1} \\
0 - 1
\end{array}
$$

Therefore, the oblique asymptote is the line $y = x + 1$.

Note: The remainder $-\dfrac{1}{x-1}$ goes to 0 as $|x| \to \infty$ so you have $x + 1$ left.

Horizontal and Oblique Asymptotes (cont.)

$$F(x) = \frac{Q(x)}{P(x)}$$

$F(x)$ is the quotient between two polynomials.

Find the horizontal or oblique asymptote.

1. $f(x) = \dfrac{(x^2 - 10x + 5)}{(3x^2 - x - 3)}$

2. $f(x) = \dfrac{(x^2 - 6x + 5)}{(x - 3)}$

3. $f(x) = \dfrac{(6x + 5)}{(10x^2 - 3x + 7)}$

4. $f(x) = \dfrac{(x^5 - 7x^2 + 5)}{(x^2 - 3x + 7)}$

5. $f(x) = \dfrac{(6x + 5)}{(x^2 - 3x + 7)}$

6. $f(x) = \dfrac{9x^2}{(3x^2 + 7)}$

Graph the Quotient of Two Polynomials

$$F(x) = \frac{Q(x)}{P(x)}$$

$F(x)$ is the quotient between two polynomials.

Example: Graph the quotient of two polynomials.

$$f(x) = \frac{-(x-1)}{(x^2 - 10x + 9)}$$

Step 1: Find the x- and y-intercepts. To find the y-intercept, set $x = 0$ and solve. $f(0) = -\frac{1}{9}$. To find the x-intercept(s), set y or $f(x)$ equal to zero and solve. There is no solution. The only intercept is $(0, +\frac{1}{9})$.

Step 2: To find the vertical asymptote(s) and/or point(s) of discontinuity, set the denominator equal to zero and solve.

$x^2 - 10x + 9 = 0$

$(x-1)(x-9) = 0, \quad x = 1$ and $x = 9$.

Because $x = 1$ makes the denominator and numerator zero, $(1, -\frac{1}{8})$ is a point of discontinuity, and $x = 9$ is a vertical asymptote.

Step 3: Take the limit of $f(x)$ as $|x|$ approaches infinity.

$$\lim_{|x| \to \infty} \frac{-(x-1)}{x^2 - 10x + 9} = 0$$

Therefore, $y = 0$ is a horizontal asymptote.

Step 4: Graph the information from steps 1, 2, and 3.

Step 5: Plot points in each region.

$x = 8, f(-3) = f(8) = 1$

$x = 10, f(10) = 1$

Step 6: Use the vertical and horizontal asymptotes, the x- and y-intercepts, the point of discontinuity, and the plotted points to sketch the graph.

Step 7: If you have a graphing calculator, use this as a check for your graph in step 6.

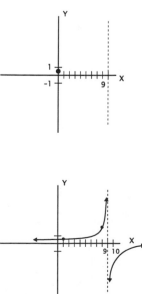

Answer Key

6. $x^4 = 16 = 2^4 \qquad x = 2$

Page 13.

1. $2x + 1 = x - 3 \qquad x = -4$

2. $2 - 2x = x + 2$

$\qquad 0 = 3x \qquad x = 0$

3. $2 = x - 4 \qquad x = 6$

4. $(3^2)^{x+7} = 3^{3-x}$

$\qquad 3^{2x+14} = 3^{3-x}$

$\qquad 2x + 14 = 3 - x$

$\qquad 3x = -11$

$\qquad x = \dfrac{-11}{3}$

Page 15.

1. $\log 5^{x+2} = \log 10^{2x}$

$\qquad (x + 2)\log 5 = 2x$

$\qquad x \log 5 + 2\log 5 = 2x$

$\qquad x \log 5 - 2x = -2\log 5$

$\qquad x = \dfrac{-2\log 5}{\log 5 - 2} = 1.074$

2. $\log 7x = \log 10^{2x-5}$

$\qquad x \log 7 = 2x - 5$

$\qquad x \log 7 - 2x = -5$

$\qquad x(\log 7 - 2) = -5$

$\qquad x = \dfrac{-5}{(\log 7 - 2)} = 4.330$

3. $\log 10^x = \log 6^{x-2}$

$\qquad x \log 10 = (x - 2)\log 6$

$\qquad x = x\log 6 - 2\log 6$

$\qquad x - x\log 6 = -2\log 6$

$\qquad x(1 - \log 6) = -2\log 6$

$\qquad x = \dfrac{-2\log 6}{1 - \log 6} \qquad x = -7.015$

4. $\log 7^{2x} = \log 3^{5x-2}$

$\qquad 2x\log 7 = (5x - 2)\log 3 = 5x\log 3 - 2\log 3$

$\qquad 2x\log 7 - 5x\log 3 = -2\log 3$

$\qquad x(2\log 7 - 5\log 3) = -2\log 3$

$\qquad x = \dfrac{-2\log 3}{(2\log 7 - 5\log 3)} = 1.372$

5. $3x\log 5 = (2x - 3)\log 8 = 2x\log 8 - 3\log 8$

$\qquad 3x\log 5 - 2x\log 8 = -3\log 8$

$\qquad x(3\log 5 - 2\log 8) = -3\log 8$

$\qquad x = \dfrac{-3\log 8}{(3\log 5 - 2\log 8)} = -9.319$

6. $2x\log 9 = (-x + 7)\log 3 = -x\log 3 + 7\log 3$

$\qquad 2x\log 9 + x\log 3 = 7\log 3$

$\qquad x(2\log 9 + \log 3) = 7\log 3$

$\qquad x = \dfrac{7\log 3}{(2\log 9 + \log 3)} = 1.400$

7. $x\log 12 = (2x - 2)\log 6 = 2x\log 6 - 2\log 6$

$\qquad x\log 12 - 2x\log 6 = -2\log 6$

$\qquad x(\log 12 - 2\log 6) = -2\log 6$

$\qquad x = \dfrac{-2\log 6}{(\log 12 - 2\log 6)} = 3.262$

8. $\qquad 2x\log 11 = (x - 1)\log 102$

$\qquad 2x\log 11 = x\log 102 - \log 102$

$\qquad 2x\log 11 - x\log 102 = -\log 102$

$\qquad x(2\log 11 - \log 102) = -\log 102$

$\qquad x = \dfrac{-\log 102}{(2\log 11 - \log 102)} = -27.075$

Page 17.

1. $\qquad 2000 = 800\left(1 + \dfrac{.035}{12}\right)^{12t}$

$\qquad \dfrac{20}{8} = \left(\dfrac{12.035}{12}\right)^{12t}$

$\qquad \log\left(\dfrac{20}{8}\right) = 12t\log\left(\dfrac{12.035}{12}\right)$

$\qquad t = \dfrac{\log 2.5}{12\log\left(\frac{12.035}{12}\right)} = 26.2$ years

2. $2500 = P\left(1 + \dfrac{.045}{6}\right)^{24} = P(1.0075)^{24}$

$\qquad P = \dfrac{2500}{1.0075^{24}} = \$2089.58$

3. $A = 2000\left(1 + \dfrac{.04}{6}\right)^{60} = \$2979.69$

Notice difference in money in the following formula. $A = 2000(1.0067)^{60} = \$2985.62$

4. $\dfrac{4000}{2000} = \dfrac{2000}{2000}\left(1 + \dfrac{.04}{6}\right)^{6t}$

$\qquad 2 = \left(\dfrac{6.04}{6}\right)$

$\qquad \log 2 = 6t\log\left(\dfrac{6.04}{6}\right)$

$\qquad t = \dfrac{\log 2}{6\log\left(\frac{6.04}{6}\right)} = 17.4$ years

Page 19.

1. $1200 = 300e^{r(4)}$

$\qquad 4 = e^{4r}$

$\qquad \ln 4 = 4r\ln e$

$\qquad r = \dfrac{\ln 4}{4} = .3465$ or 34.66%

2. $A = 2700e^{.035(3)} = 2700e^{.105} = \$2998.92$

3. $800 = Pe^{.06(10)} = Pe^{.6} \qquad \dfrac{800}{e^{.6}} = \$439.05$

4. $\qquad 1800 = 1350e^{.062t}$

$\qquad \dfrac{1800}{1350} = e^{.062t}$

$\qquad \ln\left(\dfrac{1800}{1350}\right) = .062t\,(\ln e)$

$\qquad t = \dfrac{\ln\left(\frac{1800}{1350}\right)}{.062} = 4.64$ years

Page 21.

1. $250 = 500e^{k2700}$

$\qquad .5 = e^{k2700}$

$\qquad \ln .5 = 2700k(\ln e)$

$\qquad k = \dfrac{\ln .5}{2700} = -.0003$

2. $2000 = 10e^{k(3.5)}$

$\qquad 200 = e^{k(3.5)}$

$\qquad \ln 200 = 3.5k(\ln e)$

$\qquad k = \dfrac{\ln 200}{3.5} = 1.5138$

Page 4.

1. $-4 < 5x - 1$ and $5x - 1 < 4$
$-3 < 5x$ $5x < 5$
$\dfrac{-3}{5} < x$ $x < \dfrac{5}{5}$
$> \dfrac{-3}{5}$ and $x < 1$
                     $\dfrac{-3}{5} < x < 1$

2. $2x - 8 \le -6$ or $2x - 8 \ge 6$
$2x \le 2$ $2x \ge 14$
$x \le 1$ or $x \ge 7$

3. $-(7 + x) < 5x + 10$ and $5x + 10 < 7 + x$
$-17 < 6x$ $4x < -3$
$x > \dfrac{-17}{6}$ and $x < \dfrac{-3}{4}$
                     $\dfrac{-17}{6} < x < \dfrac{-3}{4}$

4. $4 - 2x < -6$ or $4 - 2x > 6$
$-2x < -10$ $-2x > 2$
$x > 5$ or $x < -1$

Page 5.

1. $y = -2x + 7$ $x = -2y + 7$
$x - 7 = -2y$
$y = \dfrac{x}{-2} + \dfrac{7}{2}$
$f^{-1}(x) = \dfrac{-x}{2} + \dfrac{7}{2}$
$f(f^{-1}(x)) = -2\left(\dfrac{-x}{+2} + \dfrac{7}{2}\right) + 7 = +x - 7 + 7 = x$

OR

$f^{-1}(f(x)) = f^{-1}(-2x + 7) = -\dfrac{(-2x + 7)}{2} + \dfrac{7}{2} = x$

2. $y = \dfrac{1}{x} + 4$ $x = \dfrac{1}{y} + 4$
$\dfrac{x - 4}{1} = \dfrac{1}{y}$
$y = \dfrac{1}{x - 4}$
$f^{-1}(x) = \dfrac{1}{x - 4}$
$f(f^{-1}(x)) = \dfrac{1}{\frac{1}{x-4}} + 4 = \dfrac{1}{1}\left(\dfrac{x - 4}{1}\right) + 4 = x - 4 + 4 = x$

OR

$f^{-1}(f(x)) = f^{-1}\left(\dfrac{1 + 4}{x}\right) = \dfrac{1}{\frac{1}{x} + 4} - 4 = x$

Page 7.

1. ( )^6 1   6   15   20   15   6   1
2. $(x - y)^4 = x^4 - 4x^3y + 6x^2y^2 - 4xy^3 + y^4$
3. $(a - 3b)^3 = a^3 + 3a^2(-3b) + 3a(-3b)^2(-3b)^3$
             $= a^3 - 9a^2b + 27ab^2 - 27b^3$
4a. $5^6 = 15,625$ **4b.** $2^6 = 64$
4c. $20(5x)^3(2y)^3 \therefore 20 \cdot 5^3 \cdot 2^3 = 20,000$
5. $P(x) = 7(x + y)^5$

Page 8.

1. $(x^2 - 6x + 9) - 4(y^2 + 2y + 1) = -9 + 9 - 4$
$\dfrac{(x - 3)^2}{-4} - \dfrac{4(y + 1)^2}{-4} = \dfrac{-4}{-4}$
$\dfrac{(y + 1)^2}{1} - \dfrac{(x - 3)^2}{4} = 1$

2. $x^2 + 8x + 16 + y^2 - 6y + 9 = -21 + 16 + 9$
$(x + 4)^2 + (y - 3)^2 = 4$

Page 9.

1. $3x = -2x + 5$
$5x = 5$
$x = 1$ $y = 3$ $(1, 3)$

2. $-7x = 2x^2 + 3$
$0 = 2x^2 + 7x + 3 = (2x + 1)(x + 3)$
$x = -\dfrac{1}{2}$ $y = \dfrac{7}{2}$ $\left(-\dfrac{1}{2}, \dfrac{7}{2}\right)$
$x = -3$ $y = 21$ $(-3, 21)$

Page 10.

1. $x^2 = 9 - y2$
$\dfrac{(9 - y^2)}{16} + \dfrac{y^2}{9} = 1$
$\dfrac{9}{16} - \dfrac{y^2}{16} + \dfrac{y^2}{9} = 1$
$144\left(\dfrac{-y^2}{16} + \dfrac{y^2}{9}\right) = \dfrac{7}{16}$
$-9y^2 + 16y^2 = 63$
$7y^2 = 63$
$y^2 = 9$
$y = \pm 3$ $x = 0$
$(0, 3)$ $(0, -3)$

2. $x^2 = 16 - y^2$
$16 - y^2 + y^2 + 4x = 11$
$4x = -5$
$x = \dfrac{-5}{4}$
$\left(\dfrac{-5}{4}\right)^2 + y^2 = 16$
$\dfrac{25}{16} + y^2 = 16$
$y^2 = \dfrac{256}{16} - \dfrac{25}{16}$ $\left(\dfrac{-5}{4}, \dfrac{+\sqrt{231}}{4}\right)$
$y^2 = \dfrac{231}{16}$ $\left(\dfrac{-5}{4}, \dfrac{-\sqrt{231}}{4}\right)$
$y = \dfrac{\pm\sqrt{231}}{4}$

Page 11.

1. $y - 5 = 3x^2 - 9x = 3(x^2 - 3x)$
$\dfrac{243}{4} + y - 5 = 3\left(x^2 - 9x + \dfrac{81}{4}\right)$
$\left(\dfrac{9}{2}\right)^2 = \dfrac{81}{4}$
$y - \dfrac{223}{4} = 3\left(x - \dfrac{9}{2}\right)^2$

2. $y + 3 = -5x^2 + 20x = -5(x^2 - 4x)$
$-20 + y + 3 = -5(x^2 - 4x + 4)$
$y - 17 = -5(x - 2)^2$

3. $y - 3 = x^2 + 2x$
$1 + y - 3 = x^2 + 2x + 1$
$y - 2 = (x + 1)^2$

Page 12.

1. $5^2 = x$ $x = 25$ **2.** $3^7 = 2x$ $x = \dfrac{2187}{2}$
3. $4^2 = x - 1$ $x = 17$
4. $2^x = 12$
$x \ln 2 = \ln 12$
$x = \dfrac{\ln 12}{\ln 2}$ $x = 3.58$
5. $32 = 2^{3x}$
$2^5 = 2^{3x}$
$x = \dfrac{5}{3}$

Answer Key

4.

$$\underline{-2|} \quad 9 \quad -12 \quad +17 \quad +5 \quad -25$$
$$\quad -18 \quad +60 \quad -154 \quad +298$$
$$\overline{\quad 9 \quad -30 \quad +77 \quad -149 \quad +273}$$

$9x^3 - 30x^2 + 77x - 149$ is the reduced equation.

Page 54.

1.

$$n = x^2$$
$$x^2 - 5n + 4 = 0$$
$$(n-4)(n-1) = 0$$
$$n = 4,\ 1$$
$$x^2 = 4 \qquad x^2 = 1$$
$$x = \pm 2 \qquad x = \pm 1$$

2.

$$2n^2 - n - 10 = 0$$
$$(2n-5)(n+2) = 0$$
$$n = \frac{5}{2},\ -2$$
$$x^2 = \frac{5}{2} \qquad x^2 = -2$$
$$x = \pm\sqrt{\frac{5}{2}} \qquad x = \pm\sqrt{2}\,i$$

Page 55.

1.

$$(\sin x)^2 = n$$
$$n^2 - 5n + 4 = 0$$
$$(n-1)(n-4) = 0$$
$$n = 1,\ 4$$

so 4 is not in the domain of $\sin x$

$$\sin^2 x - 1 \qquad \sin^2 x = 4$$
$$\sin x = \pm 1 \qquad \sin x = \pm 2$$
$$x = \sin^{-1} \qquad \sin x \geq 1$$
$$1 = 90^\circ \text{ or } \sin^{-1} -1 = 270^\circ$$

2.

$$\cos^2 x = n$$
$$2n^2 - n - 10 = 0$$
$$(2n-5)(n+2) = 0$$
$$n = \frac{5}{2} \text{ or } n = 2$$

but $-1 \leq \cos x \leq 1$

$\therefore$ there are no solutions

3.

$$(\log x)^2 = n$$
$$n^2 - 9n + 8 = 0$$
$$(n-1)(n-8) = 0$$
$$n = 1,\ 8$$
$$(\log x)^2 = 8$$
$$\log x = \pm\sqrt{8}$$
$$x = 10^{\sqrt{8}} \text{ or } 10^{-\sqrt{8}}$$
$$(\log x)^2 = 1$$
$$\log x = \pm 1$$
$$x = 10^1 = 10$$
$$x = 10^{-1} = \frac{1}{10}$$

4. $5n^2 - 23n - 10 = 0$

$$(5n+2)(n-5) = 0$$
$$n = \frac{-2}{5},\ 5$$
$$x = \pm\sqrt{\frac{2}{5}}\,i \qquad x = \pm\sqrt{5}$$

5.

$$n = x^4 \text{ or } n = -1$$
$$n^2 - 5n - 6 = 0 \qquad x^4 \neq 1$$
$$(n-6)(n+1) = 0 \qquad x^2 = \pm\sqrt{-1} = \pm i$$
$$x^4 = 6 \qquad x = \pm\sqrt{i} \text{ or } \pm\sqrt{-i}$$
$$x^2 = \pm\sqrt{6}$$
$$x = \pm\sqrt{\sqrt{6}} \qquad x = \pm\sqrt{\sqrt{6}}\,i$$

6.

$$n = x^3$$
$$2n^2 - 5n - 12 = 0$$
$$(2n+3)(n-4) = 0$$
$$n = \frac{-3}{2},\ 4$$
$$x^3 = \frac{-3}{2} \qquad x^3 = 4$$
$$x = \sqrt[3]{\frac{-3}{2}} \qquad x = \sqrt[3]{4}$$

Plus 4 imaginary roots.

Page 56.

1.

$$\begin{array}{r} 3x^2 - 1 \qquad\qquad r = 2x - 2 \\ x - 0x - 1 \overline{)\ 3x^4 - 0x^3 - 4x^2 + 2x - 1} \\ \underline{-(3x^4 - 0x^3 - 3x^2)} \\ 0 \quad 0 - x^2 + 2x - 1 \\ \underline{-\ (x^2 - 0x + 1)} \\ 2x - 2 \end{array}$$

2.

$$\begin{array}{r} 3x^2 + x - 3 \quad r = -x \\ x^2 - x \overline{)\ 3x^4 - 2x^3 - 4x^2 + 2x} \\ \underline{-(3x^4 - 3x^3)} \\ x^3 - 4x^2 \\ \underline{-\ (x^3 - x^2)} \\ -3x^2 + 2x \\ \underline{-(-3x^2 + 3x)} \\ -x \end{array}$$

3.

$$\begin{array}{r} x^2 + 1 \\ x^2 - 2x + 8 \overline{)\ x^4 - 2x^3 + 9x^2 - 2x + 8} \\ \underline{-(x^4 - 2x^3 + 8x^2)} \\ 0 \quad 0 \quad 1x^2 - 2x + 8 \\ \underline{-(1x^2 - 2x + 8)} \\ 0 \quad 0 \quad 0 \end{array}$$

4.

$$\begin{array}{r} 2x^2 - 2/3 \qquad r = 4/3x - 53/3 \\ 3x^2 - x - 4 \overline{)\ 6x^4 - 2x^3 - 10x^2 + \quad 2x - \quad 15} \\ \underline{-(6x^4 - 2x^3 - \quad 8x^2)} \\ 0 \quad 0 - 2x^2 + \quad 2x - \quad 15 \\ \underline{-(-\ 2x^2 + 2/3x^3 + \quad 8/3)} \\ 0 \quad 4/3x - 53/3 \end{array}$$

Page 57.

1.

$$\begin{array}{r} 3x^2 - 1 \\ x^2 - 0x - 1 \overline{)\ 3x^4 - 0x^3 - 4x^2 + 2x - 1} \\ \underline{-(3x^4 - 0x^3 - 3x^2)} \\ 0 \quad 0 - x^2 + 2x - 1 \\ \underline{-(-\ x^2 - 0x + 1)} \\ 2x - 2 \end{array}$$

$P(x) = d(x)(3x^2 - 1) + 2x - 2$

Page 42.

1. $z = \sqrt{13}(\cos 30^\circ + i \sin 30)$

$z = \sqrt{13}\left(\dfrac{\sqrt{3}}{2} + \dfrac{1}{2}i\right) = \dfrac{\sqrt{39}}{2} + \dfrac{\sqrt{13}}{2}i$

2. $z = 8(\cos 60^\circ + i \sin 60^\circ)$

$z = 8\left(\dfrac{1}{2} + \dfrac{\sqrt{3}}{2}i\right) = 4 + 4\sqrt{3}i$

3. $z = 10(\cos(-135^\circ) + i \sin(-135^\circ))$

$z = 10\left(\dfrac{-\sqrt{2}}{2} + \dfrac{\sqrt{2}}{2}i\right) = -5\sqrt{2} - 5\sqrt{2}i$

4. $z = 3(\cos 120^\circ + i \sin 120^\circ)$

$z = 3\left(-\dfrac{1}{2} + \dfrac{\sqrt{3}}{2}i\right) = \dfrac{-3}{2} + \dfrac{3\sqrt{3}}{2}i$

Page 43.

1. $35(\cos 95^\circ + i \sin 95^\circ),\ \dfrac{7}{5}(\cos 25^\circ + i \sin 25^\circ)$

2. $5(\cos 180^\circ + i \sin 180^\circ),\ 20(\cos 20^\circ + i \sin 20^\circ)$

3. $3(\cos 120^\circ + i \sin 120^\circ),\ \dfrac{1}{3}(\cos -60^\circ + i \sin -60^\circ)$

4. $60(\cos 120^\circ + i \sin 120^\circ),\ \dfrac{5}{12}(\cos 30^\circ + i \sin 30^\circ)$

Page 45.

1a. $z^3 = 6^3(\cos 60^\circ + i \sin 60^\circ) = 216(\cos 60^\circ + i \sin 60^\circ)$

1b. $z^6 = 6^6(\cos 120^\circ + i \sin 120^\circ)$

$\quad = 46{,}656(\cos 120^\circ + i \sin 120^\circ)$

2a. $z^{\frac{1}{3}} = 64^{\frac{1}{3}}(\cos 16^\circ + i \sin 16^\circ) = 4(\cos 16^\circ + i \sin 16^\circ)$

$z^{\frac{1}{3}} = 64^{\frac{1}{3}}(\cos 136^\circ + i \sin 136^\circ) = 4(\cos 136^\circ + i \sin 136^\circ)$

$z^{\frac{1}{3}} = 64^{\frac{1}{3}}(\cos 256^\circ + i \sin 256^\circ) = 4(\cos 256^\circ + i \sin 256^\circ)$

2b. $z^{\frac{1}{8}} = 64^{\frac{1}{8}}(\cos 6^\circ + i \sin 6^\circ) = 1.68(\cos 6^\circ + i \sin 6^\circ)$

$z^{\frac{1}{8}} = 1.68(\cos 51^\circ + i \sin 51^\circ) = 1.68(\cos 96^\circ + i \sin 96^\circ)$

$z^{\frac{1}{8}} = 1.68(\cos 141^\circ + i \sin 141^\circ)$

$z^{\frac{1}{8}} = 1.68(\cos 186^\circ + i \sin 186^\circ)$

$z^{\frac{1}{8}} = 1.68(\cos 231^\circ + i \sin 231^\circ)$

$z^{\frac{1}{8}} = 1.68(\cos 276^\circ + i \sin 276^\circ)$

$z^{\frac{1}{8}} = 1.68(\cos 321^\circ + i \sin 321^\circ)$

2c. $z^{\frac{1}{2}} = 64^{\frac{1}{2}}(\cos 24^\circ + i \sin 24^\circ)$

$\quad = 8(\cos 24^\circ + i \sin 24^\circ) = 8(\cos 204^\circ + i \sin 204^\circ)$

Page 47.

1. $\sin A = \dfrac{3}{5}\qquad \cos A = \dfrac{4}{5}\qquad \tan A = \dfrac{3}{4}$

$\cot A = \dfrac{4}{3}\qquad \sec A = \dfrac{5}{4}\qquad \csc A = \dfrac{5}{3}$

2. $\sin A = \dfrac{5}{13}\qquad \cos A = \dfrac{12}{13}\qquad \tan A = \dfrac{5}{12}$

$\cot A = \dfrac{12}{5}\qquad \sec A = \dfrac{13}{12}\qquad \csc A = \dfrac{13}{5}$

3. $\sin B = \dfrac{15}{17}\qquad \cos B = \dfrac{8}{17}\qquad \tan B = \dfrac{15}{8}$

$\cot B = \dfrac{8}{15}\qquad \sec B = \dfrac{17}{8}\qquad \csc B = \dfrac{17}{15}$

Page 49.

1. centerline $+1$
   amplitude 1

phase shift $-\dfrac{C}{B} = \dfrac{\pi}{3}$

$x = \dfrac{\pi}{3}$

period $\dfrac{2\pi}{3}$

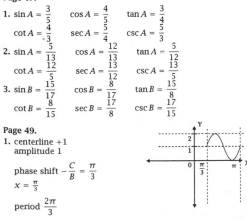

2. centerline -2

amplitude 3

phase shift $\dfrac{\pi}{2}$

period $\dfrac{2\pi}{1}$

3. centerline 2

amplitude 2

phase shift $\dfrac{\pi}{12}$

period $\dfrac{2\pi}{12}$

4. centerline 1

amplitude 3

phase shift π

period 2π

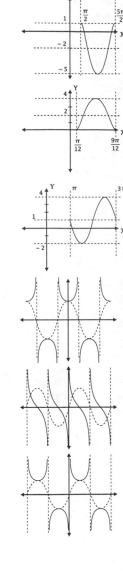

Page 51.

1. $\sec 3x = \dfrac{1}{\cos 3x}$

$y = \cos 3x$

period $\dfrac{2\pi}{3}$

2. $y = \cot 4x = \dfrac{\cos 4x}{\sin 4x}$

$y = \sin 4x$

period $\dfrac{2\pi}{4} = \dfrac{\pi}{2}$

3. $y = \csc \dfrac{x}{2} = \dfrac{1}{\sin\left(\frac{x}{2}\right)}$

$y = \sin \dfrac{x}{2}$

period $\dfrac{2\pi}{\frac{1}{2}} = 4\pi$

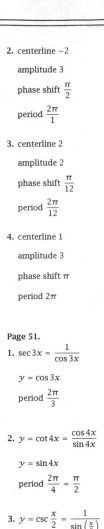

Page 53.

1.

3	5	−3	+5	+0	−10
		+15	+36	+123	+369
	5	+12	+41	+123	+359

$f(3) = 359$

2.

−2	3	−6	+2	+9	−1
		−6	+24	−52	+86
	3	−12	+26	−43	+87

1. No, -2 is not a root. Also, -2 does not satisfy the rational root theorem.

3.

2	−4	−1	+7	+3	−11
		−8	−18	−22	−38
	−4	−9	−11	−19	−49

$P(2) = -49 \quad (2, -49)$

3. $10 = 50e^{-.345t}$

$\dfrac{1}{5} = e^{-.345t}$

$\ln \dfrac{1}{5} = -.345t\,(\ln e)$

$t = \dfrac{\ln .2}{-.345} = 4.7$ years

4. $A = 100e^{(2.5)10} = 100e^{25} = 100(7.2 \times 10^{10}) = 7.2 \times 10^{12}$

Page 22.
1. $4^2 + 5^2 + 6^2 + 7^2 + 8^2 = 190$
2. $3^2 + 3^3 + 3^4 + 3^5 + 3^6 = 9 + 27 + 81 + 243 + 729 = 1089$
3. $8 + 7 + 6 + 5 + 4 + 3 = 33$

Page 23.

1.
$tn = 3 + (n-1)5 = 5n - 2 \qquad \displaystyle\sum_{n=1}^{4} 5n - 2$

2.
$tn = -5 + (n-1)4 = -5 + 4n - 4 = 4n - 9 \qquad \displaystyle\sum_{n=1}^{4} 4n - 9$

Page 24.

1. $a = 3 \qquad n = 8 \qquad r = 2$
$s_8 = \dfrac{3(1 - 2^8)}{1 - 2} = \dfrac{3(-255)}{-1} = 765$

2.
$s_5 = \dfrac{1}{2}\left(\dfrac{1 - \left(\frac{1}{2}\right)^5}{1 - \frac{1}{2}} \right) = \dfrac{1}{2} \cdot \dfrac{\left(1 - \left(\frac{1}{2}\right)^5\right)}{\frac{1}{2}}$

$= 1 - \left(\dfrac{1}{2}\right)^5 = 1 - \dfrac{1}{32} = \dfrac{31}{32}$

Page 25.

1. $S = \dfrac{9}{1 - r} = |r| < 1 \qquad r = 2$

The sum does not exist (infinite) because $|r| > 1$.

2. $S = \dfrac{9}{1 - r} \qquad r = \dfrac{1}{3}$

$S = \dfrac{\frac{3}{2}}{1 - \frac{1}{3}} = \dfrac{\frac{3}{2}}{\frac{2}{3}} = \dfrac{3}{2} \cdot \dfrac{3}{2} = \dfrac{9}{4}$

3. $S = \dfrac{9}{1 - r} \qquad r = \dfrac{-1}{3}$

$S = \dfrac{-2}{1 - \left(-\frac{1}{3}\right)} = \dfrac{-2}{\frac{4}{3}} = \dfrac{-2}{1} \cdot \dfrac{3}{4} = \dfrac{-6}{4}$

$= -1.5 \text{ or } \dfrac{-3}{2}$

Page 27.
1. Let $s = \{n \in N : a_n = 2^n + 1\}$
   1. Show $1 \in S \quad a_1 = 2^1 + 1 = 3 \quad$ true $1 \in S$
   2. Assume $x \in S \quad 9x = 2^x + 1$
     Prove $x + 1 \in S \quad a_{x+1} = 2^{x+1} + 1$
   3. Proof: $a_{x+1} = 2a_x - 1$
       $a_x = 2^x + 1$
       $a_{x+1} = 2(2^x + 1) - 1$ (substitute)
       $a_{x+1} = 2 \cdot 2^x + 2 - 1$
       $a_{x+1} = 2^{x+1} + 1 \qquad \therefore S = N$

2. Let $S = \left\{ n \in N : a_n = 5\left(\dfrac{1}{2}\right)^{n-1} \right\}$

   1. Show $1 \in S \quad a_1 = 5\left(\dfrac{1}{2}\right)^{1-1} = 5 \cdot 1 = 5 \quad$ true

   2. Assume $x \in S \quad a_x = 5\left(\dfrac{1}{2}\right)^{x-1}$

     Prove $x + 1 \in S \quad a_{x+1} = 5\left(\dfrac{1}{2}\right)^{(x+1)-1}$

   3. Proof: $a_{x+1} = \dfrac{a_x}{2}$

       $a_x = 5\left(\dfrac{1}{2}\right)^{x-1}$

       $a_{x+1} = \dfrac{5\left(\frac{1}{2}\right)^{(x-1)}}{2}$

       $= 5\left(\dfrac{1}{2}\right)^{(x-1)}\left(\dfrac{1}{2}\right)^1$ (substitute)

       $= 5\left(\dfrac{1}{2}\right)^{(x+1)-1} \qquad \therefore S = N$

3. Let $S = \left\{ n \in N : a_n = \dfrac{n}{2} - \dfrac{a}{2} \right\}$

   1. Show $1 \in S \quad a_1 = \dfrac{1}{2} - \dfrac{9}{8} = \dfrac{-8}{2} = -4$

                               true $1 \in S$

   2. Assume $x \in S \quad a_x = \dfrac{x}{2} - \dfrac{9}{2}$

     Prove $x + 1 \in S \quad a_{x+1} = \dfrac{x+1}{2} - \dfrac{9}{2}$

   Proof: $a_{x+1} = a_x + \dfrac{1}{2}$

       $a_x = \dfrac{x}{2} - \dfrac{9}{2}$

       $a_{x+1} = \dfrac{x}{2} - \dfrac{9}{2} + \dfrac{1}{2}$ (substitute) $= \left(\dfrac{x}{2} + \dfrac{1}{2}\right) - \dfrac{9}{2}$

       $= \left(\dfrac{x+1}{2}\right) - \dfrac{9}{2} \qquad \therefore S = N$

Page 29.

1. Let $S = \left\{ n \in N : \dfrac{1}{2} + \dfrac{1}{6} + \dfrac{1}{12} + \cdots + \dfrac{1}{n(n+1)} = \dfrac{n}{n+1} \right\}$

   1. Show $1 \in S \quad \dfrac{1}{2} = \dfrac{1}{1(1+1)} = \dfrac{1}{2}$ true $1 \in S$

   2. Assume $x \in S \quad \dfrac{1}{2} + \dfrac{1}{6} + \dfrac{1}{12} + \cdots + \dfrac{1}{x(x+1)} = \dfrac{x}{x+1}$

     Prove $x + 1 \in S \quad \dfrac{1}{2} + \dfrac{1}{6} + \dfrac{1}{12} + \cdots + \dfrac{1}{x(x+1)} + \dfrac{1}{(x+1)[(x+1)+1]} = \dfrac{x+1}{(x+1)+1} = \dfrac{x+1}{x+2}$

   3. Proof: $\dfrac{1}{2} + \dfrac{1}{6} + \dfrac{1}{12} + \cdots + \dfrac{1}{x(x+1)} + = \dfrac{1}{(x+[(x+1)+1]}$

       $= \dfrac{x}{x+1} + \dfrac{1}{(x+1)[(x+1)+1]}$

       $= \dfrac{x}{(x+1)} + \dfrac{(x+2)}{(x+2)} + \dfrac{1}{(x+1)(x+2)}$

       $= \dfrac{x^2 + 2x + 1}{(x+1)(x+2)} = \dfrac{(x+1)(x+1)}{(x+1)(x+2)} = \dfrac{(x+1)}{(x+2)} \qquad \therefore S = N$

Answer Key

2. Let $S = \left\{ n \in N : 1^3 + 2^3 + 3^3 + \ldots n^3 = \dfrac{[n(n+1)]^2}{4} \right\}$

  1. Show $1 \in S$ $\dfrac{[1(1+1)]^2}{4} = \dfrac{2^2}{4} = \dfrac{4}{4} = 1 = 1^3$ yes $1 \in S$

  2. Assume $x \in S$ $1^3 + 2^3 + 3^3 + \ldots x^3 = \dfrac{[x(x+1)]^2}{4}$

    Prove $x + 1 \in S$ $1^3 + 2^3 + 3^3 + \ldots x^3 + (x+1)^3 = \dfrac{[(x+1)(x+2)]^2}{4} = \dfrac{(x^2 + 3x + 2)^2}{4}$

  3. Proof: $1^3 + 2^3 + 3^3 + \ldots + x^3 + (x+1)^3 = \dfrac{[x(x+1)]^2}{4} + (x+1)^3$

$$= \dfrac{(x^2 + x)^2}{4} + x^3 + 3x^2 + 3x + 1 = \dfrac{x^4 + 2x^3 + x^2 + 4x^3 + 12x^2 + 12x + 4}{4}$$

$$= \dfrac{x^4 + 6x^3 + 13x^2 + 12x + x^4}{4} = \dfrac{(x^2 + 3x + 2)^2}{4} \quad \therefore S = N$$

Page 31.
1. 3
2. 0
3. −4
4. does not exist
5. −1
6. −1

Page 33.
1. $\dfrac{(x-1)}{(x-6)} = \dfrac{-2}{-7}$ $x \neq -1$ $(-1, \frac{2}{7})$

2. $\dfrac{(x+1)}{(x+5)} = \dfrac{6}{10}$ $x \neq 5$ $(5, \frac{3}{5})$

3. $\dfrac{(x+1)}{(x-9)} = \dfrac{-8}{-18}$ $x \neq -9$ $(-9, \frac{4}{9})$

Page 35.
1. $\displaystyle\lim_{x \to \infty} f(x) = \dfrac{1}{3}$ horizontal asymptote

$$y = \dfrac{1}{3}$$

2.
$$\begin{array}{r} x - 3 \\ x-3\overline{\smash{\big)}\,x^2 - 6x + 5} \\ \underline{-x^2 + 3x} \\ -3x + 5 \end{array}$$
$y = x - 3$ oblique asymptote

3. $\displaystyle\lim_{x \to \infty} f(x) = 0$ horizontal asymptote $y = 0$

4. $\displaystyle\lim_{x \to \infty} f(x) =$ does not exist

5. $\displaystyle\lim_{x \to \infty} f(x) = 0$ horizontal asymptote $y = 0$

6. $\displaystyle\lim_{x \to \infty} f(x) = \dfrac{9}{3} = 3$ horizontal asymptote $y = 3$

Page 37.
1. $y = 0$ horizontal asymptote

$\left(0, \dfrac{2}{9}\right) (-2, 0)$

$\dfrac{x+2}{(x-9)(x+1)}$

$x = 9, -1$ vertical asymptotes

2. $\dfrac{(x-2)(x+2)}{(x-2)(x-2)}$

$y = 1$ horizontal asymptote

$(0, -1)$ $(x^2 - 4) = 0$
$(-2, 0)$ $x = -2, 2$

$x = +2$ vertical asymptote

Note: In original equation $f(x) = \dfrac{0}{0}$, but there is no point of discontinuity because in reduced equation $g(x) = \dfrac{x+2}{x-2}$ and $g(2)$ is undefined.

3. $\dfrac{x^2}{(x-2)(x-4)}$
$y = 1$ horizontal

$x = 2, 4$ vertical

$f(3) = \dfrac{9}{9 - 18 + 8} = \dfrac{9}{1} = -9$

Page 38.
1. $\dfrac{(x+3)(x-3)}{(x+1)(x-3)}$ $x = -1$

2. $\dfrac{(x-4)(x+1)}{(x-4)(x+4)}$ $x = -4$

3. $\dfrac{(x+6)(x+1)}{(x+1)(x-1)}$ $x = 1$

4. $\dfrac{(x-2)(x-2)}{(x-2)(x+1)}$ $x = -1$

Page 39.
1. sum $14 - 6i$, difference $8 + 2i$, product $25 - 50i$
2. sum $8 + 6i$, difference $6 + 8i$, product 14
3. sum $11 + 7i$, difference $1 - 3i$, product $20 + 40i$
4. sum $8 - 6i$, difference $8 - 12i$, product $27 + 24i$
5. sum $39 + 9i$, difference $15 - 35i$, product $610 + 438i$
6. sum 20, difference $2\sqrt{2}i$, product 102

Page 40.
1. $\dfrac{(6+6i)}{(6+6i)} = \dfrac{18 + 18 - 18i + 18i}{72}$

$$\dfrac{36}{72} = \dfrac{1}{2}$$

2. $\dfrac{(1-i)}{(1-i)} = \dfrac{2 + 7 + 7i - 2i}{2} = \dfrac{9}{2} + \dfrac{5i}{2}$

3. $\dfrac{(-2i)}{(-2i)} = \dfrac{-2 - 24i}{4} = \dfrac{-1}{2} - 6i$

4. $\dfrac{(3 - 4i)}{(3 - 4i)} = \dfrac{15 + 20 + 15i - 20i}{25}$

$$\dfrac{35}{25} - \dfrac{5}{25}i = \dfrac{7}{5} - \dfrac{1}{5}i$$

Page 41.
1. $|z| = \sqrt{58}$ $\tan^{-1}\left(\dfrac{-7}{3}\right) = -66.8°$ $\left(\sqrt{58}, -66.8°\right)$

2. $|z| = \sqrt{85}$ $\tan^{-1}\left(\dfrac{6}{7}\right) = 40.6°$ $\left(\sqrt{85}, 40.6°\right)$

3. $|z| = 5\sqrt{2}$ $\theta = -45°$ $\left(5\sqrt{2}, -45°\right)$

4. $\sqrt{45} = 3\sqrt{5}$

$\tan^{-1}\left(\dfrac{3}{-6}\right) + 180° = -26.57° + 180°$ second quadrant

$\left(3\sqrt{5}, 153.4°\right)$

2.

$$\begin{array}{r} 3x^2 + x - 3 \\ x^2 - x\sqrt{3x^4 - 2x^3 - 4x^2 + 2x} \\ \underline{-(3x^4 - 3x^3)} \\ 0 \quad x^3 - 4x^2 \\ \underline{-(\ x^3 - x^2)} \\ -3x^2 + 2x \\ \underline{-(-3x^2 + 3x)} \\ -x \end{array}$$

$P(x) = d(x)(3x^2 + x - 3) - x$

3.

$$\begin{array}{r} x^2 + 1 \\ x^2 - 2x + 8\sqrt{x^4 - 2x^3 + 9x^2 - 2x + 8} \\ \underline{-(x^4 - 2x^3 + 8x^2)} \\ 0 \quad 0 + x^2 - 2x + 8 \\ \underline{-(\ x^2 - 2x + 8)} \\ 0 \quad 0 \quad 0 \end{array}$$

$P(x) = d(x)(x^2 + 1)$

4.

$$\begin{array}{r} 2x^2 - 2/3 \\ 3x^2 - x - 4\sqrt{6x^4 - 2x^3 - 10x^2 + \ 2x - \ 15} \\ \underline{-(6x^4 - 2x^3 - \ 8x^2)} \\ 0 \quad 0 - 2x^2 + \ 2x - \ 15 \\ \underline{-(- 2x^2 - 2/3x + \ 8/3)} \\ 0 \ \ 4/3x - 53/3 \end{array}$$

$P(x) = d(x)(2x^2 - 2/3) + (4/3x - 53/3)$

Page 58.

1. $\begin{array}{r} 3\rfloor \quad 4 - \ 2 + \ 6 - \ \ 5 - \ \ 19 \\ \underline{+ 12 + 30 + 108 + 309} \\ 4 + 10 + 36 + 103 + 290 \end{array}$

$P(x) = (x - 3)(4x^3 + 10x^2 + 36x + 103) + 290$

2. $\begin{array}{r} 1\rfloor \quad 1 - 4 + 7 - 2 - 9 \\ \underline{+ 1 - 3 + 4 + 2} \\ 1 - 3 + 4 + 2 - 7 \end{array}$

$P(x) = (x - 1)(x^3 - 3x^2 + 4x + 2) - 7$

3. $\begin{array}{r} -2\rfloor \quad 5 - \ 1 + \ 2 - \ \ 5 - \ \ 1 \\ \underline{- 10 + 22 - 48 + 106} \\ 5 - 11 + 24 - 53 + 105 \end{array}$

$P(x) = (x + 2)(5x^3 - 11x^2 + 24x - 53) + 105$

4. $\begin{array}{r} -4\rfloor \quad 2 - \ 3 + \ 7 - \ \ 8 - \ \ 11 \\ \underline{- 8 + 44 - 204 + 848} \\ 2 - 11 + 51 - 212 + 837 \end{array}$

$P(x) = (x + 4)(2x^3 - 11x^2 + 51x - 212) + 837$

Page 59.

1. $\pm 1, \ \pm\dfrac{1}{3}$ **2.** $\pm 1, \ \pm 3, \ \pm 9$

3. $\pm 1, \ \pm\dfrac{1}{3}, \ \pm\dfrac{5}{3}, \ \pm 5$ **4.** $\pm 1, \ \pm 3, \ \pm 9, \ \pm\dfrac{1}{5}, \ \pm\dfrac{3}{5}, \ \pm\dfrac{9}{5}$

Page 60.

1. $\begin{array}{r} 2\rfloor \quad 2 - \ 4 + \ 6 - \ 24 \\ \underline{4 \quad \ \ 0 \quad \ \ 12} \\ 2 \quad \ \ 0 \quad \ \ 6 - 12 \end{array}$
$(x - 2)$ is not a factor, $P(x) = (x - 2)(2x^2 + 6) - 12$

2. $\begin{array}{r} 5\rfloor \quad 1 - \ 4 + \ 6 - \ 25 \\ \underline{5 \quad \ \ 5 \quad \ \ 55} \\ 1 \quad \ \ 1 \quad 11 \quad \ \ 30 \end{array}$
$(x - 5)$ is not a factor, 5 is not a root
$P(x) = (x - 5)(x^2 + x + 11) + 30$

3. $\begin{array}{r} 12\rfloor \quad 1 - \ 32 + \ 64 - \ 2112 \\ \underline{12 - 240 - 2112} \\ 1 - 20 - 176 - \ \ 0 \end{array}$
yes, 12 is a factor, $P(x) = (x - 12)(x^2 - 20x - 176)$

Page 61.

1. $u \cdot v = 4(-8) + -4(10) = -72$, not perpendicular
2. $u \cdot v = 7(24) + -12(14) = 168 - 168 = 0$, perpendicular
3. $u \cdot v = 27(2) + 32(-3) = 59 - 96 = -42$, not perpendicular

4. $u \cdot v = 6(3) + -4(-2) = 26$
not perpendicular
but coincidental

5. If one vector times a
scalor equals the other
vector, then the vectors are
parallel and coincidental.
$2u = v$

Page 62.

1. $u \cdot v = 6(-2) + -3(-5) = 3$
   $|u| = \sqrt{45}$ $|v| = \sqrt{29}$
   $\theta = \cos^{-1}\left(\dfrac{3}{\sqrt{45}\cdot\sqrt{29}}\right) = 85.2°$

2. $u \cdot v = -3(-4) + (-10)(4) = -28$
   $|u| = \sqrt{109}$ $|v| = 4\sqrt{2}$
   $\theta = \cos^{-1}\left(\dfrac{-28}{\sqrt{109}\cdot 4\sqrt{2}}\right) = 118.3°$

Page 63.

1. $f(4) = -(4)^2 + 4 = -12$ $f(0) = -(0)^2 + 4 = +4$
   $\dfrac{-12 - 4}{4 - 0} = -4$ -4 is the average rate of change

2. $f(5) = \dfrac{2}{5}$ $f(i) = \dfrac{2}{1}$
   $\dfrac{\frac{2}{5} - 2}{5 - 1} = \dfrac{\frac{-8}{5}}{4} = \dfrac{-8}{5}\cdot\dfrac{1}{4} = \dfrac{-2}{5}$

3. $f(2) = (2)^4 - 3(2)^2 + 4 = 8$ $f(1) = (-1)^4 - 3(-1)^2 + 4 = 2$
   $\dfrac{8 - 2}{2 - (-1)} = \dfrac{6}{3} = 2$

Page 65.

1. $f(x + h) = (x + h)^3 - 4$
   $f'(x) = \lim\limits_{h\to 0} (x + h)^3 - 4 - (x^3 - 4)$
   $f'(x) = \lim\limits_{h\to 0} \dfrac{(x^3 + 3x^2h + 3xh^2 + h^3) - 4 - x^3 + 4}{h}$
   $f'(x) = \lim\limits_{h\to 0} \dfrac{(3x^2h + +3xh^2 + h^3)}{h}$
   $= \lim\limits_{h\to 0} \dfrac{h(3x^2 + 3xh + h^2)}{h} = 3x^2$

2. $f(x+h) = 3(x + h)^3 - 4(x + h) + 5$
   $f'(x) = \lim\limits_{h\to 0} 3(x + h)^3 - 4(x + h) + 5 - (3x^3 - 4x + 5)$
   $f'(x) = \lim\limits_{h\to 0} \dfrac{3x^3 + 9x^2h + 9xh^2 + 3h^3 - 4x - 4h + 5 - 3x^3 + 4x - 5}{h}$
   $f'(x) = \lim\limits_{h\to 0} \dfrac{9x^2h + 9xh^2 + 3h^3 - 4h}{h}$
   $= \lim\limits_{h\to 0} \dfrac{h(9x^2 + 9xh + 3h^2 - 4)}{h} = 9x^2 - 4$

3. $f(x + h) = 3(x + h)^2 - (x + h) + 27$

$$f'(x) = \lim_{h \to 0} \frac{3(x + h)^2 - (x + h) + 27 - (3x^2 - x + 27)}{h}$$

$$f'(x) = \lim_{h \to 0} \frac{3x^2 + 6xh + 3h^2 - x - h + 27 - 3x^2 - x - 27}{h}$$

$$= \lim_{h \to 0} \frac{h(6x + 3h - 1)}{h} = 6x - 1$$

Page 66.

1. $f(1 + h) = (1 + h)^3 - (1 + h)$

$$f'(1) = \lim_{h \to 0} \frac{(1 + h)^3 - (1 + h) - (1^3 - 1)}{h}$$

$$f'(1) = \lim_{h \to 0} \frac{1 + 3h + 3h^2 + h^3 - 1 - h}{h}$$

$$f'(1) = \lim_{h \to 0} \frac{h(2 + 3h + h^2)}{h} = 2$$

2. $f(-7 + h) = 2(-7 + h)^2 - 3(-7 + h)$

$$= +98 - 28h + 2h^2 + 21 - 3h = 119 - 31h + 2h^2$$

$$f'(-7) = \lim_{h \to 0} \frac{119 - 31h + 2h^2 - 119}{h}$$

$$f'(-7) = \lim_{h \to 0} \frac{-31h + 2h^2}{h} = \lim_{h \to 0} \frac{h(-31 + 2h)}{h} = -31$$

Page 67.

1. $f(2 + \triangle t) = 3(2 + \triangle t)^2 - 4(2 + \triangle t) + 8$

$f(2 + \triangle t) = 12 + 12\triangle t + 3(\triangle t)^2 - 8 - 4\triangle t + 8$

$f(2 + \triangle t) = 12 + 8\triangle t + 3\triangle t^2$

$$f'(2) = \lim_{\triangle t \to 0} \frac{12 + 8\triangle t + 3\triangle t^2 - 12}{\triangle t}$$

$$f'(2) = \lim_{\triangle t \to 0} \frac{\triangle t(8 + 3\triangle t)}{\triangle t} = 8$$

2. $f(3 + \triangle t) = 4.9(3 + \triangle t)^2 + 16(3 + \triangle t) + 10$

$f(3 + \triangle t) = 44.1 + 29.4\triangle t + 4.9(\triangle t)^2 + 48 + 16 + 10$

$f(3 + \triangle t) = 102.1 + 45.4\triangle t + 4.9(\triangle t)^2$

$$f'(3) = \lim_{\triangle t \to 0} \frac{102.1 + 45.4\triangle t + 4.9(\triangle t)^2 - 102.1}{\triangle t}$$

$$f'(3) = \lim_{\triangle t \to 0} \frac{\triangle t(45.4 + 4.9\triangle t)}{\triangle t} = 45.4$$

3. $f(10 - \triangle t) = 5(10 - \triangle t)^2 - 4 = 500 - 100\triangle t - 5(\triangle t)^2 - 4$

$$= 496 - 100\triangle t - 5(\triangle t)^2$$

$$f'(10) = \lim_{\triangle t \to 0} \frac{496 - 100\triangle t - 5(\triangle t)^2 - 496}{\triangle t}$$

$$f'(10) = \lim_{\triangle t \to 0} \frac{\triangle t(-100 - 5\triangle t)}{\triangle t} = 100$$

4. $f(3 + \triangle t) = (3 + \triangle t)^3 - 3(3 + 3\triangle t)^2 - 1$

$f(3 + \triangle t) = 27 + 27\triangle t + 9\triangle t^2 + (\triangle t)^3$

$f(3 + \triangle t) = -1 + 9\triangle t + 6(\triangle t)^2 + (\triangle t)^3$

$$f'(3) = \lim_{\triangle t \to 0} \frac{-1 + 9\triangle t + 6(\triangle t)^2 + (\triangle t)^3 + 1}{\triangle t}$$

$$f'(3) = \lim_{\triangle t \to 0} \frac{\triangle t(+9\triangle t + 6\triangle t + (\triangle t)^2)}{\triangle t} = 9$$

Page 69.

1. $f(x) = 3x^2 - 2$ is a quadratic formula.

$$a = 3 \quad b = 0$$

$f'(x) = 2(3)x + 0 = 6x$

$f'(1) = 6 \quad f(1) = 1$

$m = 6 \ (1, 1) \quad y - 1 = 6(x - 1)$

2. $f(x + h) = \dfrac{2}{x + h}$

$$f'(x) = \lim_{h \to 0} \frac{\frac{2}{x+2} - \frac{2}{x}}{h} = \lim_{h \to 0} \frac{2x - 2x - 2h}{(x + h)(x)} \cdot \frac{1}{h}$$

$$= \frac{-2}{x^2} = f'(4) = \frac{-2}{4^2} = \frac{-1}{8}$$

$$m = \frac{-1}{8} \quad f(4) = \frac{1}{2} \quad \left(4, \frac{1}{2}\right)$$

$$y - \frac{1}{8} = \frac{-1}{8}(x - 4)$$

3. $f(x + h) = (x + h)^{\frac{1}{2}}$

$$f'(x) = \lim_{h \to 0} \frac{(x + h)^{\frac{1}{2}} - x^{\frac{1}{2}}}{h} \cdot \frac{(x + h)^{\frac{1}{2}} + x^{\frac{1}{2}}}{(x + h)^{\frac{1}{2}} + x^{\frac{1}{2}}}$$

$$= \lim_{h \to 0} \frac{x + h - x}{[(x + h)^{\frac{1}{2}} + x^{\frac{1}{2}}]} = \frac{1}{2x^{\frac{1}{2}}}$$

$$f'(9) = \frac{1}{6} \quad m = \frac{1}{6} \quad f(9) = 3 \quad (9, 3)$$

$$y - 3 = \frac{1}{6}(x - 9)$$

Page 71.

1. $3x^3 - 12x^2 + 12x - 3 \qquad 9x^2 - 24x + 12$

$3(3x^2 - 8x + 4) \qquad 3(3x - 2)(x - 2) \qquad \dfrac{2}{3}, 2$

Increasing when $x < \dfrac{2}{3}$ and $x > 2$

Decreasing when $\dfrac{2}{3} < x < 2$

$y = .55 \quad y = -3$

2. $3x^2 - 6x - 45 \qquad 3(x^2 - 2x - 15)$

$3(x + 3)(x - 5) \qquad -3, 5$

Increasing when $x < -3$ and $x > 5$

Decreasing when $-3 < x < 5$

$y = 136 \quad y = -174$

3. $6x^2 - 78x + 72 \qquad 6(x^2 - 13x + 12)$

$(x - 12)(x - 1) \qquad 1, 12$

Increasing when $x < 1$ and $x > 12$

Decreasing when $1 < x < 12$

$y = 31 \quad y = -1300$

Page 72.

1. $f'(x) = 6x - 2$
2. $f'(x) = 49x^6 - 9x^2$
3. $f'(x) = 2 - 6x - 16x^3 + 7$
4. $f'(x) = -2x^{-3} + 2x$
5. $f'(x) = 4x - 3 - 2x^{-2}$

6. $f'(x) = \dfrac{-2}{(2x)^2}$

Graph the Quotient of Two Polynomials (cont.)

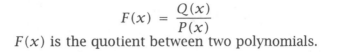

$$F(x) = \frac{Q(x)}{P(x)}$$

$F(x)$ is the quotient between two polynomials.

Graph the quotient of two polynomials.

1. $f(x) = \dfrac{(x + 2)}{(x^2 - 8x - 9)}$

2. $f(x) = \dfrac{(x^2 - 4)}{(x^2 - 4x + 4)}$

3. $f(x) = \dfrac{x^2}{(x^2 - 6x - 8)}$

Vertical Asymptotes

$$F(x) = \frac{Q(x)}{P(x)}$$
$F(x)$ is the quotient between two polynomials.

Following is the algorithm for finding the vertical asymptotes for the graph of the quotient of two polynomials.

Step 1: Set the denominator equal to zero.

Step 2: Factor and solve for x, if possible.

Step 3: Substitute the value(s) for x into the numerator of the quotient. Those values for which the numerator is not zero are vertical asymptotes.
Note: The values where the numerator is zero are points of discontinuity.

Find the vertical asymptotes.

Example:
$$p(x) = \frac{(2x - 2)}{(x^2 - 4)}$$

Step 1: $x^2 - 4 = 0$

Step 2: $(x - 2)(x + 2) = 0$; $x = -2$; and $x = 2$ are candidates for vertical asymptotes.

Step 3: The numerator is $2x - 2$ with $2(-2) - 2 = -6 \neq 0$ and $2(2) - 2 = 2 \neq 0$. Therefore, the lines $x = -2$ and $x = 2$ are both vertical asymptotes.

1. $f(x) = \dfrac{x^2 - 9}{x^2 - 2x - 3}$

2. $f(x) = \dfrac{x^2 - 3x - 4}{x^2 - 16}$

3. $f(x) = \dfrac{x^2 + 7x + 6}{x^2 - 1}$

4. $f(x) = \dfrac{x^2 - 4x + 4}{x^2 - x - 2}$

Sum, Difference, and Product

> A complex number is written $a + bi$.
> Where a and b are real numbers,
> a is called the *real part*, bi is
> called an *imaginary number*.
> ($i = \sqrt{-1}$ so $i^2 = -1$).

Example: Find the sum, difference, and product for the given complex numbers.

Given $z = 5 - 3i$ and $w = 4 + 7i$, combine the real parts and the imaginary numbers.

$$z + w = 5 - 3i + 4 + 7i$$
$$= 5 + 4 - 3i + 7i$$
$$= 9 + 4i$$
$$z - w = 5 - 3i - (4 + 7i)$$
$$= 5 - 3i - 4 - 7i$$
$$= 5 - 4 - 3i - 7i$$
$$= 1 - 10i$$

$$zw = (5 - 3i)(4 + 7i)$$
$$= 20 + 35i - 12i - 21i^2$$
$$= 20 + 23i - 21(-1)$$
$$= 20 + 21 + 23i$$
$$= 41 + 23i$$

Find the sum, difference, and product for the given complex numbers. (For the difference, subtract the second complex number from the first.)

1. $11 - 2i$ and $3 - 4i$

2. $7 + 7i$ and $1 - i$

3. $6 + 2i$ and $5 + 5i$

4. $8 - 9i$ and $0 + 3i$

5. $27 - 13i$ and $12 + 22i$

6. $10 + \sqrt{2i}$ and $10 - \sqrt{2i}$

Conjugate and Quotient

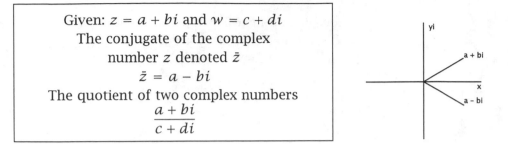

Given: $z = a + bi$ and $w = c + di$
The conjugate of the complex
number z denoted $\bar{z}$
$$\bar{z} = a - bi$$
The quotient of two complex numbers
$$\frac{a + bi}{c + di}$$

To find the quotient of two complex numbers in $a + bi$ form, multiply by one in the form of the conjugate of the numerator divided by the conjugate of the denominator and simplify.

Example: Given $z = 3 - 2i$ and $w = 4 + 5i$, find $\dfrac{z}{w}$ in complex form $(a + bi)$.

$$\frac{z}{w} = \frac{(z)(\bar{w})}{(w)(\bar{w})} = \frac{(3 - 2i)(4 - 5i)}{(4 + 5i)(4 - 5i)}$$
$$= \frac{12 - 15i - 8i + 10i^2}{16 - 25i^2}$$
$$= \frac{12 - 23i + 10(-1)}{16 - 25(-1)} = \frac{2 - 23i}{41} = \frac{2}{41} - \frac{23i}{41}$$

Find the quotient of two complex numbers in $a + bi$ form.

1. $\dfrac{3 - 3i}{6 - 6i}$

2. $\dfrac{2 + 7i}{1 + i}$

3. $\dfrac{12 - i}{2i}$

4. $\dfrac{5 + 5i}{3 + 4i}$

Convert to Polar Form

Given the complex number $z = x + yi$

The modulus or absolute value of z written $|z| = \sqrt{x^2 + y^2}$

$\theta = \tan^{-1}(\frac{y}{x})$ for first and fourth quadrant angles

$z = (|z|, \theta)$

$\theta = \tan^{-1}(\frac{y}{x}) + 180°$ for second and third quadrant angles

---

The modulus of z is also called the radius of z and is the distance from the origin to the point (x, y) on the complex plane. θ is the angle that the line containing the origin and the point (x, y) makes with the positive x-axis.

Example: Convert the complex number $z = 3 + 2i$ to polar form.

Step 1: find $|z| = \sqrt{3^2 + 2^2} = \sqrt{13}$

Step 2: Find $\theta = tan^{-1}\frac{2}{3} = 33.7°$ (first quadrant angle)

Therefore, $z = (\sqrt{13}, 33.7°)$

Convert the complex number to polar form (degrees to nearest tenth if rounding is necessary).

1. $3 - 7i$

2. $7 + 6i$

3. $5 - 5i$

4. $-6 + 3i$

Convert Polar to Trig and $a + bi$ Forms

> Given $z = (|z|, \theta)$ in polar form
> Trig form of z
> $z = r(\cos\theta + i\sin\theta)$
> $r = |z|$
> $x = r\cos\theta$ and $y = r\sin\theta$

Example: Convert the polar form of the complex number $z = (5, 45°)$ to trig form and then to $a + bi$ form.

Step 1: Substitute into the formula for trig form.
$z = 5(\cos 45° + i\sin 45°)$

Step 2: Find the $\sin 45°$ and the $\cos 45°$ and substitute.
$\sin 45° = \dfrac{\sqrt{2}}{2}$ and $\cos 45° = \dfrac{\sqrt{2}}{2}$

Step 3: Substitute into the formulas for x and y.
$x = \dfrac{5\sqrt{2}}{2}$ and $y = \dfrac{5\sqrt{2}}{2}$

Therefore, $z = 5(\cos 45° + i\sin 45°)$ in trig form and $z = \dfrac{5\sqrt{2}}{2} + \dfrac{5\sqrt{2}i}{2}$ in $a + bi$ form.

Convert the polar form of the complex number to trig form and then to $a + bi$ form. (Round to nearest hundredth.)

1. $z = (\sqrt{13}, 30°)$

2. $z = (8, 60°)$

3. $z = (10, -135°)$

4. $z = (3, 120°)$

Product and Quotient of Trig Forms

> Given $z = a(\cos\mu + i\sin\mu)$
> and $w = b(\cos\beta + i\sin\beta)$
> $zw = ab(\cos(\mu+\beta) + i\sin(\mu+\beta))$
> $\dfrac{z}{w} = \dfrac{a}{b}(\cos(\mu-\beta) + i\sin(\mu-\beta))$

Example: Find the product and the quotient of the two complex numbers.
   $z = 3(\cos 45° + i\sin 45°)$ and $w = 5(\cos 30° + i\sin 30°)$

Step 1: Substitute into the formula for the product.
   $zw = (3)(5)(\cos(45° + 30°) + i\sin(45° + 30°))$

Step 2: Evaluate the formula.
   $zw = 15(\cos 75° + i\sin 75°)$

Step 3: Substitute into the formula for the quotient.
   $z/w = .6(\cos(45° - 30°) + i\sin(45° - 30°))$

Step 4: Evaluate the formula.
   $z/w = .6(\cos 15° + i\sin 15°)$

Find the product and the quotient of the two complex numbers.

1. $7(\cos 60° + i\sin 60°)$
   and $5(\cos 35° + i\sin 35°)$

2. $10(\cos 100° + i\sin 100°)$
   and $.5(\cos 80° + i\sin 80°)$

3. $(\cos 30° + i\sin 30°)$
   and $3(\cos 90° + i\sin 90°)$

4. $5(\cos 75° + i\sin 75°)$
   and $12(\cos 45° + i\sin 45°)$

Powers and Roots: De Moivres Theorem

$$\text{Given } z = r(\cos\mu + i\sin\mu)$$
$$z^n = r^n(\cos n\mu + i\sin n\mu)$$
$$z^{1/n} = r^{1/n}\left(\cos\frac{(m+k360)}{n} + i\sin\frac{(m+k360)}{n}\right)$$
$$k \in \{0,1,2,3,4,\ldots,n-1\}$$

Example 1: Given $z = 3(\cos 30° + i\sin 30°)$, find z^5.

Step 1: Substitute $r = 3$, $\mu = 30°$, and $n = 5$ into the formula.
$$z^5 = 3^5(\cos(5)30° + i\sin(5)30°)$$

Step 2: Evaluate.
$$z^5 = 243(\cos 150° + i\sin 150°)$$

Example 2: Given $z^{1/5} = 32(\cos 30° + i\sin 30°)$, find the five complex roots of z.

Step 1: Substitute $r = 32$, $m = 30°$, and $n = 5$.
$$z^{1/5} = 32^{1/5}\left(\cos\frac{(30° + k360°)}{5} + i\sin\frac{(30° + k360°)}{5}\right)$$

Step 2: Set $k = 0$ and evaluate.
$$z^{1/5} = 32^{1/5}\left(\cos\frac{(30° + (0)360°)}{5} + i\sin\frac{(30° + (0)360°)}{5}\right)$$
$$z^{1/5} = 2\left(\cos\frac{30°}{5} + i\sin\frac{30°}{5}\right) = 2(\cos 6° + i\sin 6°)$$

Step 3: Repeat step 2 for $k = 1, 2, 3$, and 4.
$$z^{1/5} = 32^{1/5}\left(\cos\frac{(30° + (1)360°)}{5} + i\sin\frac{(30° + (1)360°)}{5}\right)$$
$$z^{1/5} = 2(\cos 78° + i\sin 78°)$$
$$z^{1/5} = 32^{1/5}\left(\cos\frac{(30° + (2)360°)}{5} + i\sin\frac{(30° + (2)360°)}{5}\right)$$
$$z^{1/5} = 2(\cos 150° + i\sin 150°)$$
$$z^{1/5} = 32^{1/5}\left(\cos\frac{(30° + (3)360°)}{5} + i\sin\frac{(30° + (3)360°)}{5}\right)$$
$$z^{1/5} = 2(\cos 222° + i\sin 222°)$$
$$z^{1/5} = 32^{1/5}\left(\cos\frac{(30° + (4)360°)}{5} + i\sin\frac{(30° + (4)360°)}{5}\right)$$
$$z^{1/5} = 2(\cos 294° + i\sin 294°)$$

Powers and Roots: De Moivres Theorem (cont.)

1. Given $z = 6(\cos 20 + i \sin 20)$, find
   a. z^3

   b. z^6

2. Given $z = 64(\cos 48 + i \sin 48°)$, find
   a. the 3 cube roots of z.

   b. the 8 eighth roots of z.

   c. the 2 square roots of z.

Triangular Trigonometry

Following are the formulas for the six trig functions in a right triangle.

$$\sin A = \frac{\text{leg opposite}}{\text{hypotenuse}} \qquad \cos A = \frac{\text{leg adjacent}}{\text{hypotenuse}} \qquad \tan A = \frac{\text{leg opposite}}{\text{leg adjacent}}$$

$$\csc A = \frac{\text{hypotenuse}}{\text{leg opposite}} \qquad \sec A = \frac{\text{hypotenuse}}{\text{leg adjacent}} \qquad \cot A = \frac{\text{leg adjacent}}{\text{leg opposite}}$$

For angle A:

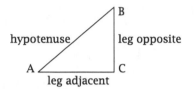

Example: Find the values of the six trig functions for angle A in the right triangle with the given measurements.

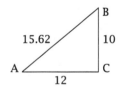

$$\sin A = \frac{10}{15.62} \qquad \cos A = \frac{12}{15.62} \qquad \tan A = \frac{10}{12}$$

$$\csc A = \frac{15.62}{10} \qquad \sec A = \frac{15.62}{12} \qquad \cot A = \frac{12}{10}$$

Triangular Trigonometry (cont.)

1. Find the values of the six trig functions for angle A in the right triangle with the given measurements.

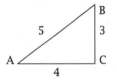

2. Find the values of the six trig functions for angle A in the right triangle with the given measurements.

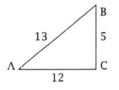

3. Find the values of the six trig functions for *angle B* in the right triangle with the given measurements.

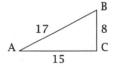

Graphing Trig Functions: Sine and Cosine

$$f(x) = A\sin(Bx + C) + D$$
$$f(x) = A\cos(Bx + C) + D$$
$$D = \text{new center line}$$
$$A = \text{amplitude}$$
$$-C/B = \text{phase shift}$$
$$2\pi/B = \text{period}$$

Graph one cycle of the trig function.

Example: $f(x) = 3\sin(2x + \pi) - 1$

Step 1: Draw the new center line, $y = -1$.

Step 2: Since the amplitude is 3, draw lines $y = 2$ and $y = -4$ above and below the center line. This will eventually help you draw the curve at the correct amplitude.

Step 3: Find the phase shift. $\dfrac{-C}{B} = \dfrac{-\pi}{2}$

Draw the line $x = \dfrac{-\pi}{2}$.

Step 4: Find the period, $p = \dfrac{2\pi}{2} = \pi$. Proceed from the phase shift line a distance of π and draw a line $x = \dfrac{\pi}{2}$.

Step 5: Draw the characteristic curve.
For sin, $f(x) = \sin x$, $\sim$, (or if cosine, $f(x) = \cos x$, $\vee$). One cycle of the curve fits inside the box drawn from steps 1 through 4.

The graph of $f(x) = 3\sin(2x + \pi) - 1$:

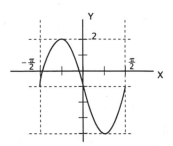

Graphing Trig Functions:
Sine and Cosine (cont.)

Graph one cycle of the given trig function.

1. $f(x) = \sin(3x - \pi) + 1$

2. $f(x) = 3\cos\left(x - \dfrac{\pi}{2}\right) - 2$

3. $f(x) = -2\cos\left(3x - \dfrac{\pi}{4}\right) + 2$ **Note:** A negative amplitude flips the curve.

4. $f(x) = -3\sin(x - \pi) + 1$

Graphing Trig Functions:
Tangent, Cotangent, Secant, and Cosecant

$$\tan x = \frac{\sin x}{\cos x} \qquad \sec x = \frac{1}{\cos x}$$

$$\cot x = \frac{\cos x}{\sin x} \qquad \csc x = \frac{1}{\sin x}$$

Characteristic curves:

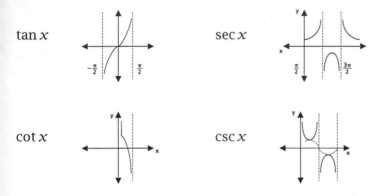

$\tan x$

$\sec x$

$\cot x$

$\csc x$

To graph one of these four trig functions, graph the denominator, locate where the denominator is zero and draw vertical asymptotes. Determine the characteristic curve and graph between the asymptotes.

Example: Graph $f(x) = \tan(2x)$.

Step 1: Graph $f(x) = \cos(2x)$. (See p. 62.)

Step 2: Where the graph crosses the x-axis locates vertical asymptotes. Draw the vertical asymptotes.

Step 3: Draw the characteristic curve.

Graphing Trig Functions:
Tangent, Cotangent, Secant, and Cosecant (cont.)

Graph the trig function.

1. $f(x) = \sec(3x)$

2. $f(x) = \cot(4x)$

3. $f(x) = \csc\left(\dfrac{x}{2}\right)$

Synthetic Substitution

Synthetic substitution is used in many ways in precalculus. It is used to find the roots and evaluate, factor, and divide polynomials. These procedures would be very time consuming without synthetic substitution.

Set up:

Step 1: Bring down the coefficients.

$f(x) = 6x^3 - 5x^2 + 7x - 8$

Step 2: Leave a space below the coefficients and draw a line.

$$\underline{1|} \quad \begin{array}{cccc} 6 & -5 & +7 & -8 \\ \hline 6 \end{array}$$

Step 3: Bring the leading coefficient below the line.

Step 4: Place the number for x that is used to evaluate $f(x)$ to the left of the coefficients (inside the symbol __|).

Evaluation:

Step 5: Multiply 1 times 6 (the divisor by the first coefficient) and place it under the -5. Then add that column. Put the result, 1, under the line.

$$\underline{1|} \quad \begin{array}{cccc} 6 & -5 & +7 & -8 \\ & 6 & +1 & +8 \\ \hline 6 & +1 & +8 & 0 \end{array}$$

Step 6: Repeat the process (multiply the divisor by the sum of the second column): 1 times 1 = 1. Place the result under the 7 and add. Then put the result, 8, under the line.

Step 7: Repeat the process until there are no more coefficients.

Note: When setting up synthetic substitution, make sure that all terms are included in the polynomial.

Example: Evaluate $p(x) = 5x^5 - 2x^3 + x - 1$ at $x = -2$

Set up:
Let $p(x) = 5x^5 + 0x^4 - 2x^3 + 0x^2 + x - 1$

$$\underline{-2|} \quad \begin{array}{cccccc} 5 & +0 & -2 & +0 & +1 & -1 \end{array}$$

Synthetic Substitution (cont.)

1. Evaluate $f(x)$ at $x = 3$.
   $f(x) = 5x^4 - 3x^3 + 5x^2 + 0x - 10$ **Note:** 0 is the coefficient of x.

2. Determine if -2 is a root of the polynomial.
   $p(x) = 3x^4 - 6x^3 + 2x^2 + 9x - 1$

3. Find the y-coordinate for the given polynomial if the x-coordinate is 2.
   $p(x) = -4x^4 - x^3 + 7x^2 + 3x - 11$

4. Find the reduced equation when the given polynomial is divided by $x = -2$.
   $p(x) = 9x^4 - 12x^3 + 17x^2 + 5x - 25$

Chunking: Converting to a Quadratic

$$f(x) = a(g(x))^4 + b(g(x))^2 + c = 0$$

Whenever there is an equation in the form shown above, it can be converted to a quadratic using the following procedure.

Example: Find all the values that make $f(x) = 0$.
$$f(x) = x^4 - 2x^2 - 3 = 0$$

Step 1: Label $x^2 = n$ and $(x^2)^2 = x^4 = n^2$ and substitute into $f(x)$.
$$f(x) = n^2 - 2n - 3 = 0$$

Step 2: Solve for n. In this case use factoring.
$$(n + 1)(n - 3) = 0 \text{ and } n = -1 \text{ or } n = 3$$

Step 3: Re-substitute x^2 in for n in the solutions in step 3.
$$x^2 = -1 \text{ and } x^2 = 3.$$

Step 4: Solve these equations for x.
$$x^2 = -1, x = \pm\sqrt{-1}, x = \pm i$$
$$x^2 = 3, x = \pm\sqrt{3}$$
Therefore, the 4 values that solve the equation are $\pm i$ and $\pm\sqrt{3}$.

Convert the equation to a quadratic and find the solutions.

1. $f(x) = x^4 - 5x^2 + 4 = 0$ 　　　　　　2. $f(x) = 2x^4 - x^2 - 10 = 0$

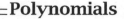

Chunking: Converting to a Quadratic (cont.)

Convert the equation to a quadratic and find the solutions. Check the answers with the original equation. (Some of the solutions obtained may not satisfy the original equation.)

1. $f(x) = \sin^4 x - 5\sin^2 x + 4 = 0$

2. $f(x) = 2\cos^4 x - \cos^2 x - 10 = 0$

3. $f(x) = (\log x)^4 - 9(\log x)^2 + 8 = 0$

4. $f(x) = 5x^4 - 23x^2 - 10 = 0$

5. $f(x) = x^8 - 5x^4 - 6 = 0$

6. $f(x) = 2x^6 - 5x^3 - 12 = 0$

Long Division of Polynomials

Long division of polynomials follows the same step as long division of rational numbers.

Example: Divide $x^4 - 4x^2 + 6x - 12$ by $x^2 - 3$

$$x^2 - 1r = 6x - 15$$
$$x^2 + 0x - 3\sqrt{x^4 + 0x^3 - 4x^2 + 6x - 12}$$
$$-(x^4 + 0x^3 - 3x^2)$$

Step 1: Set up the division and put in the zero place holders.

$$-x^2 + 6x - 12$$
$$-(-x^2 + 0x + 3)$$
$$0 + 6x - 15$$

Step 2: What must x^2 be multiplied by to give x^4? Answer x^2.

Step 3: Multiply and place in the right columns.

Step 4: Substract and bring down.

Step 5: Repeat as needed.

Perform the division.

1. Divide $3x^4 - 4x^2 + 2x - 1$ by $x^2 - 1$

2. $\dfrac{(3x^4 - 2x^3 - 4x^2 + 2x)}{(x^2 - x)}$

3. $\dfrac{(x^4 - 2x^3 + 9x^2 - 2x + 8)}{(x^2 - 2x + 8)}$

4. $\dfrac{6x^4 - 2x^3 - 10x^2 + 2x - 15}{(3x^2 - x - 4)}$

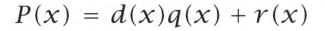

$$P(x) = d(x)q(x) + r(x)$$

$p(x) = d(x)q(x) + r(x)$
$p(x)$ is any polynomial
$d(x)$ is the divisor
$q(x)$ is the quotient
$r(x)$ is the remainder

Any polynomial can be written in the form of the divisor times the quotient plus the remainder. Here are the same exercises from the preceding page (Long Division). Write them in the form of the divisor times the quotient plus the remainder.

Perform the division. Write as $p(x) = d(x)q(x) + r(x)$.

1. $p(x) = 3x^4 - 4x^2 + 2x - 1d(x)$

  $= x^2 - 1$

  Find $q(x)$ and $r(x)$.

2. $\dfrac{(3x^4 - 2x^3 - 4x^2 + 2x)}{(x^2 - x)}$

3. $\dfrac{(x^4 - 2x^3 + 9x^2 - 2x + 8)}{(x^2 - 2x + 8)}$

4. $\dfrac{6x^4 - 2x^3 - 10x^2 + 2x - 15}{(3x^2 - x - 4)}$

The Remainder Theorem

Consider $P(x) = d(x)q(x) + r(x)$. Suppose the divisor $d(x)$ is $x - a$, where a is a constant. Then $r(x)$ will be a constant; call it R. This is true because the divisor $x - a$ is degree 1 and the remainder has a degree 1 less than the divisor or degree 0 (or the remainder is 0); then

$$P(x) = d(x)q(x) + r(x)$$
$$P(x) = (x - a)q(x) + R$$
$$P(a) = (a - a)q(a) + R = 0 + R \qquad \therefore R = P(a)$$

We now have what is called the remainder theorem: $P(x) = (x - a)q(x) + P(a)$

Example: Divide the polynomial $p(x) = 3x^4 - 7x^2 + 6x - 12$ by $(x - 2)$ and express the answer in $p(x) = (x - a)q(x) + p(a)$ form.

$$\begin{array}{r|rrrrr} 2] & 3+ & 0- & 7+ & 6- & 12 \quad \text{note } a = 2 \text{ not } -2 \\ & & 6 & 12 & 15 & 42 \\ \hline & 3 & 6 & 5 & 21 & 30 \end{array}$$

$p(2) = 30$ and the reduced equation is $q(x) = 3x^3 + 6x^2 + 5x + 21$.
Therefore, $p(x) = (x - 2)(3x^3 + 6x^2 + 5x + 21) + 30$.

Use synthetic substitution to express the polynomial as the product of the divisor and the quotient plus the remainder.

1. $\dfrac{4x^4 - 2x^3 + 6x^2 - 5x - 19}{(x - 3)}$

2. $\dfrac{x^4 - 4x^3 + 7x^2 - 2x - 9}{(x - 1)}$

3. $\dfrac{5x^4 - x^3 + 2x^2 - 5x - 1}{(x + 2)}$

4. $\dfrac{2x^4 - 3x^3 + 7x^2 - 8x - 11}{(x + 4)}$

The Rational Root Theorem

The rational root theorem states that in a polynomial with integer coefficients, if the polynomial has any rational roots, then the root's numerator must be a factor of the constant term (c) and the root's denominator must be a factor of the leading coefficient (a).

Example: Determine all the possible rational roots for the polynomial.
$$p(x) = 3x^3 - 5x^2 + 7x - 10$$

Step 1: Determine all the factors of the constant term.
Factors of 10 are $\pm 1, \pm 2, \pm 5, \pm 10$.

Step 2: Determine all the factors of the leading coefficient.
Factors of 3 are $\pm 1, \pm 3$.

Step 3: Combine all the possibilities for the numerator and the denominator.
$\pm 1, \pm 2, \pm 5, \pm 10$ and $\pm \frac{1}{3}, \pm \frac{2}{3}, \pm \frac{5}{3}, \pm \frac{10}{3}$ are all the possible rational roots.

Determine all the possible rational roots for the polynomial.

1. $p(x) = 3x^3 - 4x^2 + 3x - 1$

2. $p(x) = x^3 - 4x^2 + 8x - 9$

3. $p(x) = 3x^3 - 4x^2 + 9x - 5$

4. $p(x) = 5x^3 - 6x^2 + 2x - 9$

The Factor Theorem

The factor theorem, in general, states that when $p(r) = 0$, then r is a root of the polynomial $p(x)$, and $p(x) = (x - r)q(x)$, and the remainder is zero. $(x - r)$ is a factor of $p(x)$.

Example: Determine if $(x - 3)$ is a factor of $p(x) = x^3 - 3x^2 + 6x - 18$.
Express $p(x)$ as $p(x) = (x - r)q(x) + r(x)$.

$$\underline{3|}\ \ 1 - 3 + 6 - 18$$
$$\underline{\ \ \ \ \ \ 3\ \ \ 0\ \ \ 18}$$
$$1\ \ \ 0\ \ \ 6\ \ \ \ 0$$

Therefore, $(x - 3)$ is a factor of $p(x) = x^3 - 3x^2 + 6x - 18$, and $r(x) = 0$.
The reduced equation is $q(x) = x^2 + 0x + 6, p(x) = (x - 3)(x^2 + 6)$.

Solve.

1. Determine if $(x - 2)$ is a factor of $p(x) = 2x^3 - 4x^2 + 6x - 24$.
   Express $p(x)$ as $p(x) = (x - r)q(x) + r(x)$.

2. Determine if 5 is a root of $p(x) = x^3 - 4x^2 + 6x - 25$.
   Express $p(x)$ as $p(x) = (x - r)q(x) + r(x)$.

3. Determine if 12 is a factor of $p(x) = x^3 - 32x^2 + 64x + 2112$.

Vector Dot Product

$\mathbf{u} = (u_1, u_2)$ and $\mathbf{v} = (v_1, v_2)$
$\mathbf{u} \bullet \mathbf{v} = u_1 v_1 + u_2 v_2$
if $\mathbf{u} \bullet \mathbf{v} = 0$, then $\mathbf{u}$ is perpendicular to $\mathbf{v}$.

Example: Find the dot product for the vectors $\mathbf{u} = (3, 4)$ and $\mathbf{v} = (-8, 6)$. Is $\mathbf{u}$ perpendicular to $\mathbf{v}$?

Step 1: Substitute into the dot product formula.
$\mathbf{u} \bullet \mathbf{v} = u_1 v_1 + u_2 v_2, \mathbf{u} \bullet \mathbf{v} = 3(-8) + 4(6) = 24 - 24 = 0$
Therefore, $\mathbf{u} \bullet \mathbf{v} = 0$ and $\mathbf{u}$ is perpendicular to $\mathbf{v}$.

Find the dot product for the vectors $\mathbf{u}$ and $\mathbf{v}$. Indicate whether $\mathbf{u}$ is perpendicular to $\mathbf{v}$.

1. $\mathbf{u} = (4, -4)$ and $\mathbf{v} = (-8, 10)$.

2. $\mathbf{u} = (7, -12)$ and $\mathbf{v} = (24, 14)$.

3. $\mathbf{u} = (27, 32)$ and $\mathbf{v} = (2, -3)$.

4. Find the dot product for the vectors $\mathbf{u}$ and $\mathbf{v}$. Indicate whether $\mathbf{u}$ is perpendicular to $\mathbf{v}$. Graph the two vectors.
$\mathbf{u} = (6, -4)$ and $\mathbf{v} = (3, -2)$.

5. Graph the two vectors. Is there any conclusion from problems 4 and 5?
$\mathbf{u} = (1, -2)$ and $\mathbf{v} = (2, -4)$.

The Angle Between Two Vectors

$$\mathbf{u} = (u_1, u_2) \text{ and } \mathbf{v} = (v_1, v_2)$$
$$\mathbf{u} \bullet \mathbf{v} = u_1 v_1 + u_2 v_2$$
$$\cos \theta = \frac{\mathbf{u} \bullet \mathbf{v}}{|\mathbf{u}||\mathbf{v}|}$$
$$|\mathbf{u}| = \sqrt{(u_1)^2 + (u_2)^2}$$

To find the angle between two vectors, take the dot product between the vectors and divide it by the product of their norms. Then take the inverse cosine.

Example: Find the angle between the vectors $\mathbf{u} = (3, 7)$ and $\mathbf{v} = (-2, 5)$.

Step 1: Find the dot product.
$$\mathbf{u} \bullet \mathbf{v} = 3(-2) + 7(5) = -6 + 35 = 29$$

Step 2: Find the norms of $\mathbf{u}$ and $\mathbf{v}$ and their product.
$$|\mathbf{u}| = \sqrt{(3)^2 + (7)^2} = \sqrt{58}$$
$$|\mathbf{v}| = \sqrt{(-2)^2 + (5)^2} = \sqrt{29}$$
$$|\mathbf{u}||\mathbf{v}| = \sqrt{58} \cdot \sqrt{29}$$

Step 3: Find $\dfrac{\mathbf{u} \bullet \mathbf{v}}{|\mathbf{u}||\mathbf{v}|} = \dfrac{29}{\sqrt{58} \cdot \sqrt{29}}$

Step 4: Evaluate with a calculator the inverse cosine of $\dfrac{29}{\sqrt{58} \cdot \sqrt{29}}$.

$$\cos^{-1}(29/41) = 45°$$
Therefore, the angle between the two vectors is $45°$.

Find the angle between the vectors (to the nearest tenth of a degree).

1. $\mathbf{u} = (6, -3)$ and $\mathbf{v} = (-2, -5)$.

2. $\mathbf{u} = (-3, -10)$ and $\mathbf{v} = (-4, 4)$.

Average Rate of Change and Average Velocity

The average rate of change from $x = a$ to $x = b$ for a function $f(x)$ is the slope of the line containing the points $(a, f(a))$ and $(b, f(b))$.

Formula for average rate of change: $\dfrac{f(b) - f(a)}{b - a}, b \neq a.$

Example: Find the average rate of change for the function $f(x) = 3x^2 - 2$ from $x = 3$ to $x = 5$.

Step 1: Find $f(5) = 3(5)^2 - 2 = 73$

Step 2: Find $f(3) = 3(3)^2 - 2 = 25$

Step 3: Evaluate formula $\dfrac{73 - 25}{5 - 3} = \dfrac{48}{2} = 24$

Therefore, 24 is the average rate of change of the function from 3 to 5.

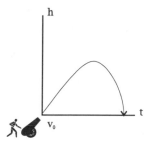

Find the average rate of change for the function given from $x = a$ to $x = b$.

1. $f(x) = -x^2 + 4 \quad a = 0 \quad b = 4$

2. $f(x) = 2/x \quad a = 1 \quad b = 5$

3. $f(x) = x^4 - 3x^2 + 4 \quad a = -1 \quad b = 2$

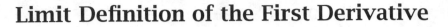
Limit Definition of the First Derivative

$$f'(x) = \lim_{h \to 0} \frac{f(x+h) - f(x)}{h}$$

The key to finding the first derivative of a function using the limit definition is to work with it algebraically until direct substitution of 0 into the formula does not result in the limit being 0/0.

Example: Using the limit definition, find the first derivative of the function
$f(x) = x^2 - x + 4$.

Step 1: Find $f(x + h)$
$$f(x + h) = (x + h)^2 - (x + h) + 4$$

Step 2: Substitute into the formula for the first derivative.
$$f'(x) = \lim_{h \to 0} \frac{(x + h)^2 - (x + h) + 4 - (x^2 + x + 4)}{h}$$

Step 3: Expand the binomial raised to a power and distribute the negative.
$$f'(x) = \lim_{h \to 0} \frac{x^2 + 2hx + h^2 - x - h + 4 - x^2 + x - 4}{h}$$

Step 4: Cancel and simplify.
$$f'(x) = \lim_{h \to 0} \frac{2hx + h^2 - h}{h}$$

Step 5: Factor out the h in the numerator and cancel with the h in the denominator.
$$f'(x) = \lim_{h \to 0} \frac{h(2x + h - 1)}{h}$$

Step 6: With the cancelling of the h, direct substitution into the limit gives us the first derivative.
$$f'(x) = 2x - 1$$

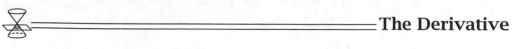

Limit Definition of the First Derivative (cont.)

Find the first derivative of the function using the limit definition.

1. $f(x) = x^3 - 4$

2. $f(x) = 3x^3 - 4x + 5$

3. $f(x) = 3x^2 - x + 27$

Instantaneous Rate of Change

$$f'(x) \text{ evaluated at a point } x = a.$$
$$f'(a) = \lim_{h \to 0} \frac{f(a + h) - f(a)}{h}$$

Example: Find the instantaneous rate of change by evaluating the first derivative at the point $x = 3$ for the function $f(x) = 3x^2 - 2x + 1$.

Step 1: Find $f(a + h)$. Expand and simplify.
$f(3 + h) = 3(3 + h)^2 - 2(3 + h) + 1$
$f(3 + h) = 3(9 + 6h + h^2) - 6 - 2h + 1$
$f(3 + h) = 27 + 18h + 3h^2 - 6 - 2h + 1$
$f(3 + h) = 3h^2 + 16h + 22$

Step 2: Find $f(3)$.
$f(3) = 3(3)^2 - 2(3) + 1 = 22$

Step 3: Substitute the results from steps 1 and 2 into the formula for the limit definition of the first derivative.
$$f'(3) = \lim_{h \to 0} \frac{3h^2 + 16h + 22 - 22}{h}$$

Step 4: Simplify, factor out an h in the numerator and cancel.
$$f'(3) = \lim_{h \to 0} \frac{h(3h + 16)}{h}$$

Step 5: Evaluate the limit by direct substitution.
$f'(3) = 16$

Therefore, the instantaneous rate of change of $f(x)$ at $x = 3$ is 16.

Find the instantaneous rate of change for the function at the given point.

1. $f(x) = x^3 - x$ at $x = 1$

2. $f(x) = 2x^2 - 3x$ at $x = -7$

Instantaneous Velocity

$$f'(t) = \lim_{\triangle t \to 0} \frac{f(t + \triangle t) - f(t)}{\triangle t}$$

A special case of the instantaneous rate of change of a function (preceding page) is when the distance is expressed as a function of time(*t*) and the first derivative evaluated at some time $t = a$ is the instantaneous velocity. In this case we generally use $\triangle t$ instead of h.

Use the steps in the example on the preceding page to find the instantaneous velocity of the function at the given time $t = a$.

1. $f(t) = 3t^2 - 4t + 8$ at $t = 2$

2. $f(t) = 4.9t^2 + 16t + 10$ at $t = 3$

3. $f(t) = 5t^2 - 4$ at $t = 10$

4. $f(t) = t^3 - 3t^2 - 1$ at $t = 3$

Tangent to the Curve at a Point

Use the limit definition of the first derivative to find the general equation, as a function of x, of the slope of the tangent line at any point. Then substitute in the given value of x and write the equation for the tangent line using point/slope form of the line.

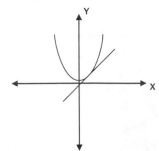

Limit definition of the first derivative

$$f'(x) = \lim_{h \to 0} \frac{f(x+h) - f(x)}{h}$$

Point/slope form of the line $\quad y - y_1 = m(x - x_1)$

Example: Find the tangent to the curve $f(x) = 3x^2 - 6x + 2$ at $x = 2$. Substitute into the limit definition of the first derivative.

$$f'(x) = \lim_{h \to 0} \frac{3(x+h)^2 - 6(x+h) + 2 - (3x^2 - 6x + 2)}{h}$$

Expand.

$$f'(x) = \lim_{h \to 0} \frac{3x^2 + 6xh + 3h^2 - 6x - 6h + 2 - 3x^2 + 6x - 2)}{h}$$

Cancel.

$$f'(x) = \lim_{h \to 0} \frac{6xh + 3h^2 - 6h}{h}$$

Factor out the h.
$$f'(x) = \lim_{h \to 0} \frac{h(6x + 3h - 6)}{h}$$

Cancel and take the limit as h approaches 0.
$$f'(x) = \lim_{h \to 0} 6x + 3h - 6 = 6x - 6$$

Substitute 2 into $f'(x)$ to find the slope and into $f(x)$ to find the y-coordinate of the point.
$$f'(2) = 6(2) - 6 = 6$$
$$f(2) = 3(2)^2 - 6(2) + 2 = 2, x_1 = 2, \text{ and } y_1 = 2.$$

The slope $m = 6$ and the point $(x_1, y_1) = (2, 2)$. Therefore, $y - 2 = 6(x - 2)$ is the tangent to the curve at the point.

Tangent to the Curve at a Point (cont.)

Find the tangent to the curve at the point.

1. Use $f(x) = 3x^2 - 2$ and $x = 1$.

2. Use $f(x) = \dfrac{2}{x}$ at $x = 4$.

3. Use $f(x) = (x)^{1/2}$ at $x = 9$.

Horizontal Tangents, $f(x)$
Increasing or Decreasing

If $f'(a) = 0$, then at the point $(a, f(a))$ there is a horizontal tangent.
If $f'(x) < 0$, then $f(x)$ is decreasing.
If $f'(x) > 0$, then $f(x)$ is increasing.

To find the horizontal tangent(s) to a curve, use the $n - 1$ rule to obtain the first derivative. Set $f'(x) = 0$ and solve for x. The root(s) of the first derivative are where the horizontal tangent(s) occur. The function evaluated at the root(s) of the first derivative is a horizontal tangent.

Example: Find the horizontal tangent(s) to the curve. Determine the intervals on the x-axis where the function is increasing or decreasing.
$$f(x) = 2x^3 - 3x^2 - 36x + 1$$

Step 1: Use the $n - 1$ rule to find $f'(x)$.
$$f'(x) = 6x^2 - 6x - 36$$

Step 2: Set $f'(x) = 0$ and solve for x.
$$6x^2 - 6x - 36 = 0$$
$$6(x^2 - x - 6) = 0$$
$6(x - 3)(x + 2) = 0$, $x = 3$ and $x = -2$ are where the horizontal tangents occur.

Step 3: Evaluate the functions at $x = 3$ and $x = -2$.
$f(3) = 2(3)^3 - 3(3)^2 - 36(3) + 1 = -107$ and
$f(-2) = 2(-2)^3 - 3(-2)^2 - 36(-2) + 1 = 43$
Therefore, $y = -107$ and $y = 43$ are horizontal tangents.

Step 4: Determine the intervals on the x-axis where the function is increasing or decreasing. Divide the x-axis into intervals using the x values of the horizontal tangents as end points of the intervals.
$x < -2, -2 < x < 3, x > 3$

Step 5. Then evaluate $f'(x)$ at any value inside the intervals.
Choose $x = -4$. $f'(-4) = 6(-4)^2 - 6(-4) - 36 = 84, f'(x) > 0$
Choose $x = 0$. $f'(0) = -36, f'(x) < 0$
Choose $x = 4$. $f'(4) = 36, f'(x) > 0$
Therefore, $f(x)$ is increasing on $x < -2$ and $x > 3$, and $f(x)$ is decreasing on $-2 < x < 3$.

Horizontal Tangents, $f(x)$
Increasing or Decreasing (cont.)

Find the horizontal tangent(s) to the curve. Determine the intervals on the x-axis where the function is increasing or decreasing.

1. $f(x) = 3x^3 - 12x^2 + 12x - 3$

2. $f(x) = x^3 - 3x^2 - 45x + 1$

3. $f(x) = 2x^3 - 39x^2 + 72x - 4$

The Power Rule or The N-Minus-One Rule

If $f(x) = ax^n$, then $f'(x) = anx^{n-1}$

This formula is called the n-minus-one rule for finding the first derivative of a polynomial function of x.

Example 1: Use the n-minus-one rule to find the first derivative of the given function.
$f(x) = 3x^2, a = 3, n = 2$. Therefore, $f'(x) = (3)(2)x^{2-1} = 6x$.

Example 2: $f(x) = 8x$, so $f'(x) = 8$ because $8x = 8x^1$, and using the $n - 1$ rule, you get $8 \cdot 1x^{1-1} = 8x^0 = 8$

Example 3: The n-minus-one rule can be used on each term in a polynomial function to find the first derivative.
$f(x) = 5x^4 - 2x^3 - 5x^2 + 8x + 11$
$f'(x) = 20x^3 - 6x^2 - 10x + 8$. (The derivative of a constant is $= 0$ because $11 = 11x^0$, and using the $n - 1$ rule, you get $11.0x^{0-1} = 0$)

Use the n-minus-one rule to find the first derivative of the polynomial functions.

1. $f(x) = 3x^2 - 2x + 5$

2. $f(x) = 7x^7 - 3x^3 - 9$

3. $f(x) = 2x - 3x^2 - 4x^4 + 7x$

4. $f(x) = x^{-2} + x^2$

5. $f(x) = 2x^2 - 3x - 10 + 2x^{-1}$

6. $f(x) = 1/(2x)$